Green Infrastructure and Public Health

There is a growing body of knowledge revealing a sweeping array of connections between public health and green infrastructure—but not until now have the links between them been brought together in one comprehensive book. *Green Infrastructure and Public Health* provides an overview of current research and theories of the ecological relationships and mechanisms by which the environment influences human health and health behaviour.

Covering a broad spectrum of contemporary understanding, Coutts outlines:

- public health models that explicitly promote the importance of the environment to health
- ways in which the quality of the landscape is tied to health
- challenges of maintaining viable landscapes amidst a rapidly changing global environment.

The book presents the case for fundamental human dependence on the natural environment and creates a bridge between contemporary science on the structure and form of a healthy landscape and the myriad ways that a healthy landscape supports healthy human beings. It presents ideal reading for students and practitioners of landscape architecture, urban design, planning, and health studies.

Christopher Coutts holds undergraduate and graduate degrees in public health and a PhD in Urban, Technological, and Environmental Planning. His research examining green infrastructure and health linkages has appeared in both peer-reviewed urban planning and public health outlets. He is an associate professor of urban and regional planning and a research associate in the Center for Demography and Population Health at Florida State University, USA.

A splendid book. Christopher Coutts extends the tradition of Aldo Leopold, Rachel Carson, and Tony McMichael, showing how green infrastructure—from sweeping natural landscapes to trees to small urban greenspaces—not only provide precious ecosystem services, but are essential to our health, well-being, and even our survival as a species. This book is evidence-based, nuanced, wise, and readable. A signal contribution to public health, environmental studies, and the design professions.
Howard Frumkin, Dean, School of Public Health, University of Washington

We all know intuitively, possibly as a result of millennia of evolution of the human psyche, that nature; local, regional and global, is good for our health. But today that is not enough to win arguments either of health promotion or for nature protection. With that in mind, this is the book that this trans-disciplinary field has been waiting for. The study of the complex interrelationships between green infrastructure and health field is attracting critical research interest from many senior research teams internationally. I think this field is on the cusp of a profound leap forward. This book contributes by providing a comprehensive and especially coherent tour of the concepts, the evidence and the questions we all need to ask. It should be on the reading list of everyone involved in the fundamental issue of how to support population health and human flourishing on an increasingly urbanised planet. Conversely it needs to be read by policy-makers and practitioners fighting to protect our fragmented urban and peri-urban eco-systems. Let's hope it stimulates new conversations and brings these two groups into a powerful alliance for change.
Marcus Grant CMLI FFPH, Environmental Stewardship for Health, Expert Advisor to the WHO European Healthy Cities Network

Christopher Coutts creatively synthesizes some powerful ways of thinking about the fundamental dependence of human health on the natural environment. By aligning the contemporary discussions of green infrastructure, ecosystem services and ecological conceptions of health, he provides a useful basis for understanding how the health of the public can improve through efforts to reduce negative human impacts on the natural environment across spatial scales. He brings together a wealth of empirical evidence from diverse areas of scientific inquiry and professional practice in making his case, while taking care to point out limitations of current knowledge. Infused with both concern and optimism about the future, the book indicates how change in harmful human practices now can engender a wide range of co-benefits for future generations. I endorse his view that, by protecting the natural environment and green infrastructure, we are essentially protecting ourselves and our health.
Terry Hartig, Ph.D., M.P.H., Professor of Environmental Psychology, Uppsala University

Green Infrastructure and Public Health

Christopher Coutts

Routledge
Taylor & Francis Group

LONDON AND NEW YORK

First published 2016
by Routledge
2 Park Square, Milton Park, Abingdon, Oxon OX14 4RN

and by Routledge
711 Third Avenue, New York, NY 10017

Routledge is an imprint of the Taylor & Francis Group, an informa business

British Library Cataloguing-in-Publication Data
A catalogue record for this book is available from the British Library

Library of Congress Cataloging in Publication Data
Coutts, Christopher, author.
Green infrastructure and public health / Christopher Coutts.
p. ; cm.
Includes bibliographical references and index.
I. Title.
[DNLM: 1. Environment Design. 2. Public Health. 3. Conservation of Natural Resources. 4. Environmental Health. 5. Health Promotion. WA 30]
RA427
362.1--dc23
2015029499

ISBN: 978-0-415-71135-7 (hbk)
ISBN: 978-0-415-71136-4 (pbk)
ISBN: 978-1-315-64762-3 (ebk)

Typeset in Bembo
by Fakenham Prepress Solutions, Fakenham, Norfolk NR21 8NN

Contents

Contents

Contents

Figures

Tables

Acknowledgments

I would first like to acknowledge the diligence and patience of the contributors to this book. It was truly an honor to work with Thomas Fischer, Micah Hahn, Ian Mell, and Salvador del Saz Salazar, all wonderful scholars and, more importantly, kind people. Terry Hartig has also been an inspiration not only through his decades of meticulous nature and health research, but also for selflessly agreeing to review a portion of this book. I would also like to acknowledge Richard Jackson who has been a crusader in advancing the prominence of the physical environment in public health. He gave a talk to a small audience in Las Cruces, New Mexico, in 1999 that I was fortunate to attend. This talk was the tipping point that set me on the path to pursuing a career devoted to exploring the role of the natural environment on health and health behaviors. I am also deeply grateful to the fantastically supportive faculty of Urban and Regional Planning at Florida State University. My colleagues allowed me to disappear on a sabbatical to focus on this book. I am proud to be part of such a wonderful collection of scholars. Louise Fox and Sade Lee at Routledge were also fantastic in quickly and thoroughly answering the many questions directed at them from the many far-flung places I found myself writing. Routledge is lucky to have them. Lastly, Christienne Stauffer Coutts has displayed nothing but support on the long road of education and nothing but tolerance in forgiving the isolation I often write in.

Contributors

Thomas B. Fischer, PhD, FIEMA, is a Professor in the School of Environmental Sciences, University of Liverpool, UK. His research focuses on various decision-making impact assessment tools. He has worked in consultancy, public administration, and academia for 25 years.

Micah Hahn, PhD, MPH, is a postdoctoral fellow at the Centers for Disease Control and Prevention/National Center for Atmospheric Research. She performed her graduate work in Epidemiology/Environment and Resources at the University of Wisconsin-Madison and her graduate work in Public Health at Emory University, USA.

Ian Mell, PhD, is a Lecturer in Planning & Civic Design at the University of Liverpool, UK. His research examines the links between green infrastructure policy, practice, and value through evaluations of greenspace investment in the UK, Europe, USA, India, and China.

Salvador del Saz Salazar, PhD, is a Professor of Environmental and Resource Economics at the Universitat de València, Spain. His main research interest is the economic valuation of the environmental goods using stated preference methods.

Introduction

In the period extending from when our most distant hominid ancestors stood up in the savannah approximately five million years ago to 10,000 years ago at the end of the last ice age, human survival and health was determined exclusively by the availability of products reaped directly from the natural environment. During this period, there was no considerable manipulation of the landscape in order to secure and increase the bounty of nature's goods and services. The basic elements for survival, allowing for evolutionary success (i.e. reproduction), were acquired exclusively from the direct reaping of the services of nature's ecological and biophysical systems. Today, despite the many technologies that may give contemporary humans a false sense of dominion over and independence from the natural environment, our health and survival are still inextricably linked to the services of nature. This misconception of independence is understandable since many humans are increasingly separated from exposure to nature's physical features and processes as daily reminders of the human dependence on nature. Technological advances have generated enormous benefits to human health and longevity for segments of humanity (often at the expense of others), but even these advances, without exception, have exploited the services of nature essential for human survival and development (Liu, Dietz, Carpenter, & Folke, 2007). Shelter, food, transportation, and, yes, even one's iPad, are created from products of nature. Our ever expanding impact on the natural environment, caused by our continued success as a species and ability to replicate ourselves in ever expanding numbers, has been so extensive that it has led to the popular, but not formally adopted, categorization of a new geological epoch, the Anthropocene (Schrag, 2012).

The technologies of the Anthropocene have thus far allowed humans to defy global environmental catastrophe and the doomsday predictions of the natural environment's inability to continue to support humans (Ehrlich, 1968), but increasingly sophisticated evaluation methods of nature's capacity to support human activity have made it clear that we are now, and have been for many decades, overextending the earth's ability to sustain the delivery of ecosystem services (Rees & Wackernagel, 1996). "Ironically, just as our collective land-use practices are degrading ecological conditions across the globe, humanity has become dependent on an ever increasing share of the biosphere's resources" (Foley et al., 2005, p. 570). At current levels of natural resource use, the earth's *natural capital* will eventually fall to a level where the basic ecosystem services on which human health and life depend will be severely compromised if not eventually exhausted. When natural capital is diminished, human health suffers. We cannot alter many of the forces of nature to which even slight alterations would result in life ceasing immediately (e.g. the gravity that maintains our orbit), but we can and have dramatically altered the terrestrial and atmospheric systems on which life and health depend. The ecological and biophysical systems not created by humans are the ones on which health is *most* dependent, and maintaining the integrity of these systems requires maintaining the landscape structure (green infrastructure) that plays a critical role in the ecology of these systems.

The fundamental connection between human health and ecosystem services has not gone unnoticed, and human health has been increasingly integrated into the array of benefits of ecosystem services (e.g. The National Research Council of the National Academies, 2013). The now decades-old warnings of the diminishing state of the environment and ecological threat (Brundtland, 1987) are slowly pushing public health to consider the natural environment and processes, recognizing that actions aimed at sustainable development are actions that are necessary for protecting and promoting public health (Institute of Medicine of the National Academies, 2013). Despite this, public health still tends to view the natural environment "with ambivalence" (Hartig, Mitchell, de Vries, & Frumkin, 2014, p. 21.2). A continued ambivalence towards the fundamental importance of the natural environment to human health is ignoring the fundamental conditions necessary for life. All other public health initiatives require that these conditions are met, and these conditions are absolutely dependent upon the natural environment. Theoretical models that attempt to represent the many external influences on human health clearly identify the natural environment as fundamental to health.

Despite this, public health is rarely involved in landscape conservation even though there is no greater basic good that could be done for long-term and sustained public health than the conservation of the landscape and the ecosystem services it supports. From Tokyo to Malawi, what humans require to survive and thrive is dependent on the natural environment and green infrastructure. There can be no health, nor humans, without it.

Green infrastructure's sweeping effect on many aspects of health make it an undeniably essential part of global and local efforts aimed at protecting human health and well-being. Green infrastructure is a necessary component for health-promoting natural systems to function, but it is also the context for many health-promoting behaviors. By protecting the natural environment and green infrastructure, we are essentially protecting ourselves and our health.

Definitions

Before proceeding further, let us first define the terms *health, nature, natural environment, representations of nature*, and *green infrastructure*. The most often cited and globally recognized definition of *health* is offered by the World Health Organization (WHO) as "a state of complete physical, mental and social well-being and not merely the absence of disease or infirmity" (WHO, 1946). Health, defined in this way, has been "honored in repetition, but rarely in application" (Evans & Stoddart, 1990, p. 1347). This is evident in the thousands of research articles produced annually examining individual and population health. Most studies of health, and particularly medicine, do not truly capture health, but rather measure exactly what the WHO definition makes clear health is not: the absence (or presence) of disease. Granted, the absence of disease is often, but not always, a prerequisite for the larger goal of a complete state of well-being.

As will be presented throughout this book, there are historic anecdotes and an increasing amount of empirical work revealing the importance of the natural environment to physical, mental, and social health. Some of these studies focus on outcomes best classified as clinically diagnosable disease or infirmity, but there are also studies that reveal the importance of the natural environment to subjective well-being. There have been proposals to include measures of well-being currently missing from the dozens of national and global measures of health (Smith, Case, Smith, Harwell, & Summers, 2013). The added measures of well-being include health (as it is typically measured: physical and mental illnesses) but also health as the WHO defines it. Taken together in an ecological approach to health, the extant evidence supports what most would accept as a tacit understanding of the fundamental importance of natural environment in supporting health as the WHO defines it; we see that the natural environment is vital not only to prevent and treat disease but also to foster physical, mental, and social well-being. This is how health is conceived in this book. We are a part and product of nature, and the presence of, and access and exposure to, the natural environment is accompanied by improved health.

For the same reason that the comprehensiveness of the WHO definition of health makes it difficult to operationalize in research, it is also unwieldy for setting any policy that is distinctly health-oriented (Evans & Stoddart, 1990). This is because striving for complete physical, mental, and social well-being is a proper and laudable objective of the policies that guide human activity, but the complex and interdependent ecological determinants of health quickly make all human activities inextricably, yet more accurately, linked to health. In the case of the natural environment, all policies that have a potential impact on the natural environment could, and should, be considered health policies.

Nature is defined as the "physical features and processes of nonhuman origin that people ordinarily can perceive, including the 'living nature' of flora and fauna, together with still and running water, qualities of air and weather, and the landscapes that comprise these and show the influence of geological processes" (Hartig et al., 2014, p. 21.2). The term nature has at times been used interchangeably with "natural environment." The visible examples of nature provided in the above definition of nature might appear to affirm their interchangeability, but I do make a distinction here because there are other forms of nature not external to people. For example, the organisms living in the microbiome of the human body could be considered nature because (1) they have physical properties, and (2) they are not a human creation. These organisms rely on humans, and we rely on them, but we did not create them.

The term *natural environment* is used here to refer to the whole of nature external to the individual and often delineated by spatial boundaries. Defining the natural environment in this way emphasizes a focus on nature "external to the individual" (i.e. in one's environment). Since the focus of this book is on green infrastructure (defined in just a moment), the term natural environment is used most often to capture nature and natural processes evident in vegetated physical spaces. These spaces include larger landscapes and also smaller greenspaces to the admitted and conscientious exclusion of other forms of nature (e.g. oceans) but always keeping in mind that what is included and excluded are dependent on one another.

The term natural environment is also used to make a distinction between it and the built environment, although these two types of environments overlap greatly with the built environment being dependent on the natural environment to sustain life. Elements of nature pervade the built environment from the often adulterated nature found in urban parks to the flora and soils that are continually trying to reclaim the built environment. While urban green infrastructure is not the focus of our examination of the health benefits of green infrastructure, urban environments come up quite often because of the need to reconcile the natural and built environments. Elements of the natural environment, whether occurring naturally (e.g. the foliage that overtakes a vacant lot) or inserted by humans (e.g. street trees), will be referred to collectively here as *representations of nature*. Both naturally occurring and inserted representations of nature are part of the natural environment because they are not made by humans, and nature and health research has revealed that both have benefits to human health.

Furthermore, the natural environment here is not used synonymously with wilderness. There are two reasons for this. First, undisturbed and pristine wilderness no longer exists because human actions with global effects have left no part of earth untouched by human influence. There are still wild places with little or no human habitation in which one can feel a sense of escape, but even these environments have been touched by human actions with global reach. For example, even the wilds of Antarctica are being altered by air pollution. Second, the overwhelming majority of the world's population does not live in such places but rather in the towns, villages, and cities where the natural environment has been severely manipulated or largely subsumed. The natural environment is the environment in which all other types of environments (e.g. urban, human–social) are nested.

The natural environment and representations of nature are taken together here in the term *green infrastructure* (GI). Green infrastructure has been defined as "an interconnected network of greenspace that conserves natural ecosystem values and functions and provides associated benefits to human populations" (Benedict & McMahon, 2002). Green infrastructure is a systems way of thinking about how the totality of components of the natural environment work together to provide the ecosystem services on which humans depend. The definition provided by Benedict and McMahon (2002) is expanded here beyond "greenspaces" to include many other representations of nature that are not greenspace but that certainly contribute to the naturalness of the environment. Representations of nature are considered part of GI as they aim to restore and replicate the more natural conditions that existed prior to human manipulation.

About this book

This book can serve as a primer for those wanting to learn how GI supports human life and health. For those already immersed in any one area of the subject, this book may prove useful in situating any given area of expertise within the many other ways that GI supports health and how any one area is likely influenced and complemented by others, in an ecological fashion, with subsequent health co-benefits. This review of the natural processes that support life may be the first time in a long while that many of us have revisited these principles. For all of us, the consolidation of this information makes this book an accessible reference for elucidating interrelationships and the multiple overlapping health functions of GI. This book also provides a review of the theory of why GI is a fundamental public health issue and, in doing so, makes it clear that public health alone cannot address the public health issue of GI degradation. Urban and environmental planners, geographers, environmental scientists, to name but a few, all play a role in protecting the life-supporting and health-enhancing services of GI.

The purpose of this book is not to provide guidance on gaps in research and the methodological advances needed to address the shortcomings of our current state of knowledge. This has been done several times in the extant literature—most notably in a book on this very topic (Ward Thompson, Aspinall, & Bell, 2010) and in a recent review of the state of the science (Hartig et al., 2014). A little on research needs is covered in Chapter 10 in order to make the reader aware of these needs. This book is also not a guide to implementing GI, but rather why doing so is important to health.[1]

Admittedly, the majority of research presented in subsequent chapters is Euro- and American-centric. This is a shortcoming in that the subjective and socially constructed experience with GI certainly has an influence on its potential health benefits. There is a great need for subjective experiences to be more fully understood in a wider variety of contexts. The focus of this book is on the health benefits of GI to humans as a whole and not on the plurality of health benefits to any one demographically distinct group. There are many health benefits derived from GI from simply being human. With that said, experiences with GI do vary, and necessary distinctions between subsets of the population are made in a number of chapters.

This book answers the appeals for greater appreciation of the natural systems that support life and how they are changing (e.g. McMichael, 1993) and the appeal to jettison the reductionist approach to environmental health that ignores larger systemic threats (Patz, 2014). This is done by taking a step further back "upstream" away from proximate (micro/interpersonal) level influences on health in favor of a focus on the intermediate (meso/community) and foundational (macro) levels (Northridge, Sclar, & Biswas, 2003; Schulz & Northridge, 2004). The evidence presented in Part II of this book[2] often does not report incidence or prevalence of a specific disease associated with alterations to GI, but rather how GI affects the environmental conditions, proximate level stressors, and health behaviors that already have well-established connections to morbidity, premature mortality, and health and well-being. This upstream approach is often focused on how the fundamental level of the *natural environment* influences proximate level *agents or stressors* and not how agents or stressors influence disease. Based on the ability of GI to remove potential stressors and promote healthy behaviors at this level, the conclusion is extended that health (both the absence of disease and heightened well-being) will be improved. For example, we review the essential role of GI in the hydrological cycle and to the provisioning ecosystem service of clean and abundant water without expanding on why water is essential to health; we review how the cultural ecosystem service of physical activity is supported by access to GI with only a quick review of the physical and mental health conditions that regular physical activity can prevent.

The ecological models of health presented in Chapter 2 leave us with many possible pathways by which the natural environment can be connected to human health outcomes. In each of the chapters in Part II of this book—Essential ecosystem services, The challenge of climate change, Infectious disease ecology, Physical activity, Mental health, and Social capital—the health benefits of GI can be categorized as being derived from the presence of, or access and exposure to, GI. Some health-supporting ecosystem services come simply from the presence of GI (e.g. water, air, food, medicine, climate regulation), others from access to GI (e.g. physical activity, social capital), and yet others from mere exposure to GI (e.g. stress reduction). In all instances, the focus of these chapters is on how GI supports these ecosystem services essential to achieving health.

Notes

1. For a good introduction into the tools and institutions to implement green infrastructure, see Beatley (2011), Chapter 5; Benedict and McMahon (2002); Rouse and Bunster-Ossa (2013); Erickson (2006); and Johnston and Newton (2004).
2. Great efforts were taken to provide a comprehensive accounting of the state of our knowledge on the connection between GI and health. A number of studies not found in existing reviews of the literature are referenced here, but there is likely important work being done that was unintentionally overlooked. The author would be most grateful for any leads to research that was overlooked. Some publications were intentionally omitted to prevent redundancy with results confirmed in more recent studies.

References

Beatley, T. (2011). *Biophilic cities: Integrating nature into urban design and planning.* Washington, DC: Island Press.

Benedict, M. A., & McMahon, E. T. (2002). Green infrastructure: Smart conservation for the 21st century. *Renewable Resources Journal, 20*(3), 12–17.

Brundtland, G. (1987). *Our common future: Address to the World Commission on Environment and Development.* Tokyo.

Ehrlich, P. R. (1968). *The population bomb.* New York: Ballantine Books.

Erickson, D. (2006). *MetroGreen: Connecting open space in North American cities.* Washington, DC: Island Press.

Evans, R., & Stoddart, G. (1990). Producing health, consuming health care. *Social Science & Medicine, 31*(12), 1347–63.

Foley, J., DeFries, R., Asner, G., Barford, C., Bonan, G., Carpenter, S. R., et al. (2005). Global consequences of land use. *Science, 309*(5734), 570–4.

Hartig, T., Mitchell, R., de Vries, S., & Frumkin, H. (2014). Nature and health. *Annual Review of Public Health, 35*, 21.1–21.22.

Institute of Medicine of the National Academies. (2013). *Global development goals and linkages to health and sustainability.* Washington, DC: The National Academies Press.

Johnston, J., & Newton, J. (2004). *Building green: A guide to using plants on roofs, walls and pavements.* London: Greater London Authority.

Liu, J., Dietz, T., Carpenter, S., & Folke, C. (2007). Coupled human and natural systems. *Ambio, 36*(8), 639–49.

McMichael, A. J. (1993). *Planetary overload: Global environmental change and the health of the human species*. Cambridge: Cambridge University Press.

Northridge, M. E., Sclar, E. D., & Biswas, P. (2003). Sorting out the connections between the built environment and health: A conceptual framework for navigating pathways and planning healthy cities. *Journal of Urban Health, 80*(4), 556–68.

Patz, J. A. (2014). In memoriam Tony McMichael: A giant in the field of global environmental health. *EcoHealth, 11*, 449–50.

Rees, W., & Wackernagel, M. (1996). Urban ecological footprints: Why cities cannot be sustainable—and why they are a key to sustainability. *Environmental Impact Assessment Review, 16*, 223–48.

Rouse, D. C., & Bunster-Ossa, I. F. (2013). *Green infrastructure: A landscape approach*. Chicago, IL: APA Planners Press.

Schrag, D. (2012). Geobiology of the Anthropocene. In A. Knoll, D. Canfield, & K. Konhauser (Eds.), *Fundamentals of geobiology* (pp. 425–36). Chichester: John Wiley.

Schulz, A., & Northridge, M. E. (2004). Social determinants of health: Implications for environmental health promotion. *Health Education & Behavior, 31*(4), 455–71.

Smith, L. M., Case, J. L., Smith, H. M., Harwell, L. C., & Summers, J. K. (2013). Relating ecosystem services to domains of human well-being: Foundation for a U.S. index. *Ecological Indicators, 28*, 79–90.

The National Research Council of the National Academies. (2013). *Urban forestry: Toward an ecosystem services research agenda*. Washington, DC: The National Academies Press.

Ward Thompson, C., Aspinall, P., & Bell, S. (Eds.). (2010). *Innovative approaches to researching landscape and health: Open Space: People Space 2*. Abingdon: Routledge.

WHO. (1946). Preamble to the Constitution of the World Health Organization as adopted by the International Health Conference. In *Official Records of the World Health Organization, no. 2* (Vol. 2013, p. 100). Geneva: WHO.

PART I

CONNECTING GREEN INFRASTRUCTURE AND HEALTH

Chapter One

Green infrastructure, ecosystem services, and the study of their contribution to health

An advancement in our understanding of what constitutes a viable natural environment is the attention now given to how components of the natural environment work in concert in a system of green infrastructure (GI). Green infrastructure, existing in its component parts that can range in scale from a community garden to a transnational reserve, is essential to producing the ecosystem services critical to sustaining life and supporting human health. Considering that the natural environment is fundamental to life and health and that a GI approach is vital to a viable natural environment, conserving GI should be viewed as a basic and fundamental form of health promotion.

There are many historic examples of the recognized importance of the natural environment in the creation of healthy human habitats. In the nineteenth century, when urban planning and public health were inseparable functions aimed at improving the human condition in urban environments, the reintroduction of the natural environment was considered essential to addressing urban ills. The significance of the environment in public health promotion has waxed and waned, but current ecological models of health that prominently feature the physical environment represent a reawakening in public health research and practice to the importance of the environment, in general, and the natural environment, in particular. This reawakening, fueled and affirmed by a steady stream of research, is driving larger considerations of the fundamental necessity of the natural environment to global health. The natural environment is essential to delivering the basic elements of life, and it is only after these basic elements are secure that higher level health needs can be addressed.

This chapter proceeds by first defining GI and categorizing the ecosystem services that rely on GI. This is followed by a brief look at the methods by which the value of GI to health might be captured. Following this is a body of research that could be broadly classified as biophilic epidemiology. This work examines the connection between GI and general health and well-being. It is worth recalling from time to time that this is the goal of public health; it is not solely the absence of disease that allows public health to achieve this goal. Lastly, I discuss the consilience of disciplines necessary to investigate and bring to the fore the relationship between the natural environment and health. This consilience is not only necessary to investigate and understand this relationship but also to take action to protect the natural environment and health.

Green infrastructure

No single park, no matter how large and well designed, would provide the citizens with the beneficial influences of nature; instead parks need to be linked to one another and to surrounding residential neighborhoods.

Frederick Law Olmsted (as cited by Little, 1989)[1]

Neighborhood parks, national parks, parkways, forests, community gardens, and the myriad other forms of conserved private and public greenspaces, taken together and considered as a system, are what constitute a community's GI. At the very heart of the definition of GI adopted here, "an interconnected network of greenspace that conserves natural ecosystem values and functions and provides associated benefits to human populations" (Benedict & McMahon, 2002), are the benefits that the natural environment provides to humans. A focus on the benefits of GI to humans makes this definition somewhat distinct from the landscape ecology approach taken by others (Ahern, 2007) that focuses on the environmental benefits of GI as a landscape design strategy (Wright, 2011). Adopting the Benedict and McMahon (2002) definition in no way discounts the environmental benefits of GI. Rather, it acknowledges that doing right by the environment and implementing GI results in human benefits. Among the highly intertwined environmental, social, and economic benefits of GI are the health benefits associated with protecting GI as our "natural life-support system" (Benedict & McMahon, 2006, p. 1).[2]

What may typically come to mind when considering the "infrastructure" of human settlements are the roads, water and sewer pipes, and electrical conduits that together compose our *gray* infrastructure (Figures 1.1a–d). Gray infrastructure provides the physical environment supports essential to performing the daily functions that can enhance health. It is the imposition of continually expanding human-made gray infrastructure and the manipulation of the natural environment, from sea walls to irrigation ditches, that has allowed humans to proliferate.

Figures 1.1a–d: Gray infrastructure

Sources: a) Sigfrid Lundberg; b) amira_a; c) Jeremiah John McBride; d) Andrew Krupp.

There would be far fewer humans on earth if we were still relying on what nature could deliver or absorb without the environmental support of gray infrastructure delivering water and food, removing waste, and improving communication, among other things. We have devised complex systems of gray infrastructure that have allowed us to manipulate natural resources and dramatically increase longevity and improve health and quality of life, but our reliance on the gray infrastructure has understandably instilled a sense of separation from and dominion over the natural environment. Fueling our gray infrastructure has often come at the expense of conserving our GI, when, in reality, it is the green that makes the gray possible.[3] The natural resources produced by our GI are not only used to create and construct the gray, but the gray is only functional and useful when fueled by and delivering the products of the natural environment.

A focus on the gray without a consideration of the green poses a significant threat to health, but so does too much green without the gray. In no small part caused by the wants and needs of those living in urban areas putting great pressure on the GI outside of cities, there are very few people on earth who can live exclusively off the bounty of the natural environment—clean and abundant water drawn from a spring or stream and food and medicines gathered from the forest or sea—without some level of environmental manipulation. It is the need to manage the risks to life and health associated with living off the land, and the perceived "higher-level" opportunities afforded when they are managed, that has driven the majority of humans to migrate to urban and urbanizing environments. In these environments there exists varying levels of the social and physical infrastructure that can deliver the natural resources needed to maintain life and health—for as long as the green can support the functioning of the gray. Granted, the formal gray infrastructure in many of the world's largest cities is grossly inadequate to ensure optimal health, but even the millions in the most depressed of urban slums largely void of formal gray infrastructure find ways to appropriate the natural resources needed to survive. An ability to overcome in *no way* suggests that those living in these conditions do not need improved gray infrastructure. Rather, it is only to emphasize the fact that there are many different forms of physical infrastructure, at times not necessarily gray. All forms manipulate nature in some way to support life and health, and all influence and are dependent on GI to continue to function.

Human-made physical infrastructure supports human activities and health in our increasingly urbanized world where more people are occupying less space. While the accompanying reduced person-to-land-area ratio of urbanization might, on the surface, appear to be a good thing for GI conservation, the heightened demand for scarce land in urban and urbanizing environments still places enormous demands on GI. These demands cannot continue to be met without some consideration of GI conservation. Again, the gray is dependent on the green (Figure 1.2).

Let us now examine the components of GI and the principles involved in maintaining its quality. A wonderfully accessible primer on landscape ecology principles, *Landscape Ecology Principles in Landscape Architecture and Land-Use*

Figure 1.2: Green and gray infrastructure, Singapore
Source: William Cho.

Planning (Dramstad, Olson, & Forman, 1996),[4] presents the three interdependent characteristics of the living system of a landscape: structure, functioning, and change. The structure of the landscape is composed of the parks and forests as the component parts or pieces of the puzzle that, taken together, make up a GI system. The functioning of ecosystems is dependent upon the interconnected landscape structure of GI. The relationship between structure and function is continually changing. These changes are normal in dynamic natural systems but are dramatically accelerated by human alterations to the landscape. Of course, since humans are a product of nature, human alterations to the landscape could be considered natural. Even so, the pace and volume of this natural manipulation of landscape structure has resulted in compromised functioning, which can lead to an inability of the landscape to support health.

The components used to describe the structure of the landscape are the patch, corridor, and matrix. If you will, imagine an aerial view of the landscape from an airplane window. The sometimes symmetrical but also odd shaped forested or greenspaces, often surrounded by or adjacent to lands manipulated by humans, are the landscape patches. Elongated, narrower patches that may run along a river, road, or coast are corridors. Ideally, these corridors connect patches to one another. The inventory of spatially distributed patches and corridors, bounded by the scale of the assessment of the landscape, makes up the landscape matrix. A higher quality matrix consists of a connected structure of patches and corridors. Conversely, a matrix of lower quality consists of isolated patches. In urbanized areas, the built environment often cuts off small patches of greenspace from one another. Patches and corridors can be considered the components of GI, while the matrix can be considered the larger GI system.

Figures 1.3 and 1.4 represent the relationship between patches, also called hubs or nodes, and corridors, also called connectors, that comprise a landscape or GI matrix. Figure 1.3 is an abstraction of a GI matrix comprised of interconnected patches and corridors connected to larger scale regional GI components and matrices. Figure 1.4 is a photograph of an area just outside Oslo, Norway. Here we can see forested patches connected by a forested corridor. In many cases, GI patches are not connected to one another. While any GI has health benefits, no matter how small and disconnected, connected GI is what is necessary to enhance its health-supporting potential. Similar to how gray infrastructure must be connected to function as a system, so too must GI.

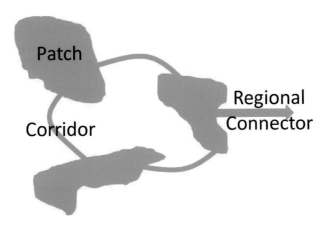

Figure 1.3: Abstraction of green infrastructure
Source: Author.

Figure 1.4: Connected green infrastructure outside Oslo, Norway
Source: Wilhelm Joys Andersen.

The bounds of any inventory of the patches and corridors that make up a GI matrix can be done at a very fine to a very broad scale. A fine scale might be an assessment of one's neighborhood environment where the parks, vegetated plazas, green roofs, abandoned lots, and river trails combine to form a localized matrix. A broader scale might consist of a metropolitan region or a state or national inventory of conserved lands. The broader scale would encompass the GI at the finer scale but it would also include larger national forests and preserves, regional parks, uncultivated agricultural lands, greenbelts, and vegetated river corridors that traverse finer scale municipal boundaries. Mapping GI assets is essential to determine how connected the system is and also how this fundamental form of infrastructure overlays and intersects other infrastructure and land uses. Similar to how Harnik and Kimball (2005) point out in their assessment of park use, "if you don't count, your park won't count," not having an inventory of GI will allow it to be undervalued in development plans. An assessment of the state of GI is essential to know where connections can be made to improve the GI matrix. Roadways are one example of gray infrastructure that can be greened to create connections. Green street networks can link greenspace at the community and regional landscape scale to mitigate the effect roads have on landscape fragmentation (Lovell & Johnston, 2009).

The scale at which GI is assessed can either mask or reveal the potential health benefits derived from the *presence* of, and *access* and *exposure* to, GI (Coutts & Horner, 2015).[5] First, there are health benefits derived simply from the *presence* of GI in one's environment. Without direct access to or even visual contact with GI, GI is working to deliver the water, food, and air on which life depends. Scale is essential here because it is unlikely that any fine scale analysis (e.g. of the neighborhood) will capture enough of the larger GI system critical to delivering these services. Instead, it is likely that the regional, and even global, scale is the appropriate scale to examine the GI components necessary to maintain a viable regional or global ecosystem. For example, the presence of the expansive, yet fleeting, South American rainforest is critical to the health of humans around the world. The health benefits derived from *access* are more localized, and a finer scale analysis, at times even smaller than the city scale, is appropriate (Richardson et al., 2012). Access denotes use, and accessing GI is much more likely if it is close to where humans live, work, and play. The benefits derived from access could range from walking into the jungle to gather food to using urban GI for passive and active recreation. The implications of accessibility are examined in greater detail in Chapter 6 when considering the health behavior of physical activity. Access, by default, provides *exposure*, but it is not a prerequisite for it. The mental health benefits from viewing nature are most appropriately examined at a finer scale. Everyday exposure can occur within one's home, workplace, or neighborhood or even on the commute to work.

Table 1.1 presents a number of GI components at various scales. From an individual building to the global ecosystem, the presence of, and access and exposure to, these components individually and collectively play a role in supporting public health.

In addition to scale, the health co-benefits of GI are also evident in the six GI planning and design principles of multifunctionality, connectivity, habitability, resiliency, identity, and return on investment (Rouse & Bunster-Ossa, 2013).[6]

Table 1.1: Green infrastructure components at various scales

Scale	Green infrastructure components
Building	Green roofs, gardens
	Indoor plants
	Green courtyards, atria
	Edible, native landscaping
Street	Parkways
	Vegetated swales
	Vegetated permeable surfaces
Community	Neighborhood gardens
	Neighborhood parks
	Riparian buffers
	Botanical gardens
	Urban farms
	Vacant lots
Region	Riparian systems
	Metropolitan greenbelts
	National parks/reserves
Globe	International ecological reserves

Source: Adapted from Beatley (2008), Table 17.1, p. 278. Beatley gave credit to the work of Girling and Kellett (2005).

While Rouse and Bunster-Ossa (2013) only explicitly note the principle of *habitability* as having particular relevance to health, the other planning and design principles of GI also have some implications for health and healthy behavior. *Multifunctionality* captures not only the many ecological functions of GI but also how these functions are important for human health. For example, an urban park protects the health of city residents by sequestering carbon, protecting biodiversity, and providing flood control. *Connectivity* of GI patches and corridors to one another and to the existing pedestrian infrastructure creates networks that are more likely to connect people to the places they need or want to go. Connected GI is therefore more likely to be used for active transportation and recreation. *Habitability* refers to the potential of a space to support recreation and community activities. These types of activities can result in the civic interactions that foster social capital and community *identity*. Creating a habitable environment for people, as well as other fauna and flora, includes satisfying needs as basic as water and air but also the recreational opportunities that foster well-being. The *resiliency* of the landscape makes it better able to continue to deliver benefits to humans in the face of environmental variability. An explicit recognition of human health as intertwined with these principles and an important *return on investment* has significant potential to create greater leverage for GI conservation and sustainable development more broadly. The various goods and opportunities that GI provides to humans and that support human life and health are referred to as ecosystem services.

Ecosystem services

An ecosystem is defined as "a biological community of interacting organisms and their physical environment" ("Ecosystem," 2013). Despite the terseness of this definition, it encapsulates an extremely complex set of interactions where organisms are interacting with one another, organisms are affecting their environment, and the environment, in turn, is affecting the organisms that inhabit it. Among this triad of potential interactions, the focus of this book is the influence of the physical environment on human organisms. Ecosystem services are the benefits that humans obtain from the ecosystems that GI supports, and the focus here is on the ecosystem services that GI provides to humans to sustain and enhance their health, well-being, and survival. "Nature's goods and services are the ultimate foundations of life and health" (Millennium Ecosystem Assessment, 2005a, p. iii).

The admitted anthropocentric focus here on ecosystem services in no way discounts the other two forms of interaction: the influence of humans on the environment and the interaction between organisms. Staying focused on GI, the local and global community of human organisms can have a positive influence on the physical environment through GI conservation. By degrading GI, humans are not only doing harm to other organisms and the functioning of ecosystems, they are harming themselves and their health. Thinking ecologically, the natural environment and humans are inseparable. As for the interaction between organisms, this part of the triad is captured in Chapters 5 and 8 when infectious disease and social capital, respectively, are discussed. With infectious disease, we examine the role of GI in determining the interactions between humans and non-human organisms. With social capital, we examine the benefits when GI facilitates humans interacting with one another. These are both ecological examinations as we aim to understand how these interactions are facilitated by GI. In this respect, GI could be considered as providing ecosystem services as it aids in improving health by reducing infectious disease and supporting social interactions.

The litany of ecosystem services can be organized into four categories: provisioning, regulating, cultural, and supporting (Melillo & Sala, 2008; Millennium Ecosystem Assessment, 2005a).[7] Figure 1.5 portrays the interdependence of these domains of services and how their delivery translates into the conditions necessary for health.

The *provisioning* services are likely the first to come to mind when considering the products of nature essential for health. These include the water produced as a service of the hydrological cycle, the plant and animal materials used as food and to make clothing, and the natural resources (e.g. wood, coal, oil, sun, wind) used to produce energy. These are the services provided by nature that have been exploited by humans for millennia. Indeed, it is these services that have supported human existence. *Regulating* services are those necessary for our sustained habitation of the earth, such as the purification of water as it migrates through the soil. These services also include climate regulation, carbon sequestration, flood control, biological regulation of infectious disease, and the soil fertility and pollination necessary for food production, among others. These services are the ones most likely taken for granted by most humans—the hidden services that are essential to the continued quality and abundance of many provisioning services. *Cultural* services encompass the non-material benefits of nature. These benefits include those obtained from recreating in nature, the economic benefits generated from

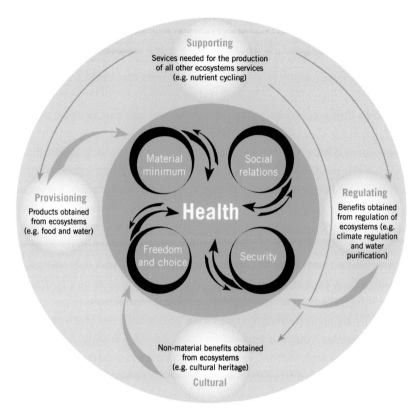

Figure 1.5: Interrelationships of categories of ecosystem services
Source: World Health Organization.

people visiting a national park, and the aesthetic and spiritual experience some feel when observing or being immersed in the natural environment.[8] *Supporting* services are those services that are necessary to produce all other ecosystem services. These are services such as soil formation and nutrient and water cycling on which the provisioning, regulating, and cultural services are dependent. Supporting services have also been called *habitat* services. The term habitat brings to the fore the significance of the landscape in a species' lifecycle and the biodiversity necessary to maintain resilient ecosystems (de Groot, Alkemade, Braat, Hein, & Willemen, 2010; de Groot, Fisher, & Christie, 2010; The Economics of Ecosystems and Biodiversity, n.d.). Green infrastructure creates the habitat essential for non-human organisms as well and supports the role they play in producing the ecosystem services on which humans depend.

Biodiversity is essential to our health because ecosystem services depend on biodiversity. Maintaining biodiversity is dependent on the protection and restoration of habitat that can support species diversity and interactions (de Groot, Wilson, & Boumans, 2002; Millennium Ecosystem Assessment, 2005b; Niemelä et al., 2010). It is quite simply "essential, not optional, to our lives and health and to our continuing to flourish as a species" (Beatley, 2011, p. 18). Green infrastructure creates the landscape structure needed to support species diversity and facilitate species interactions. The pollination of crops (discussed in Chapter 3) is a stark example of the ecosystem services that are lost when GI is degraded and biodiversity lost.

In *Sustaining Life: How Human Health Depends on Biodiversity*, Chivian and Bernstein (2008) provide an impressive compilation of the various ways in which human health is tied to biodiversity. This book contains a wealth of information on many of the topics covered here including air, water, food, and medicine, and it should be referenced if further detail is needed.

The Economics of Ecosystems and Biodiversity group provides a highly accessible accounting of ecosystem services divided by the four categories of provisioning, regulating, cultural, and habitat (or supporting) services: www.teebweb.org/resources/ecosystem-services/

The ecosystem services pertinent to health that will be covered in subsequent chapters are outlined in Table 1.2.[9, 10] This is far from an exhaustive list of ecosystem services. What this list does contain are the services where there is evidence connecting GI to the health benefits of these services. There are many, more distal benefits; for example, the potential health benefits derived from the economic development brought about by ecotourism. It is difficult to categorize the services in Table 1.2 as distinct due to the overlap of many categories, namely between the regulating and provisioning services. For example, the provisioning service of food depends greatly on the regulating services of soil fertility and pollination; the provisioning service of medicine depends on the habitat or supporting service of biodiversity.

Table 1.2: Ecosystem services

Category of service	Service
Provisioning	Water
	Food
	Medicine
Regulating	Air
	Infectious disease control
	Climate: heat, weather events
Cultural	Physical activity
	Mental health
	Social capital
Habitat or supporting	Biodiversity

Source: Author.

The US Environmental Protection Agency Eco-Health Relationship Browser (Jackson, Daniel, McCorkle, Sears, & Bush, 2013) is an interactive tool that allows users to navigate through the many connections between a number of unique ecosystems, the human health outcomes that could result from disruptions to them, and the literature to support these relationships (www. epa.gov/research/healthscience/browser/introduction.html). The continued improvement of tools such as this will make them increasingly helpful in communicating the far-reaching and interdependent effects of the loss of GI, ecological disruption, and health.

Ecosystem services can also be organized according to Abraham Maslow's enduring hierarchy of needs (1943). This involves the basic needs of food, water, etc., but it also involves higher level needs derived from the interpersonal relationships that have the potential to be initiated and fostered in accessible GI. The palimpsest of Maslow's hierarchy of needs is evident in Figure 1.6 depicting the health gradient (Barton & Tsourou, 2000; Laughlin & Black, 1995). Here we view these needs through a GI lens. Moving from the bottom to the top of the gradient, GI supports the most basic of provisioning and regulating ecosystem services needed to satisfy physical needs. Moving one step up, GI also supports the cultural ecosystem services that provide deep-seeded mental needs. Next, the gradient raises the interesting potential of GI to contribute to health knowledge. Among the many environmental and social justifications for the conservation of GI is its ability to act as a living laboratory. Observing nature and natural processes increases our knowledge of how ecosystems services are affected by human impacts on natural systems. These impacts could include the degradation of ecosystem services that occurs when pollutants are introduced into regulatory cycles (nutrient, hydrological) or the loss of the provisioning service of plants with potential medicinal properties when biodiversity is compromised.

Moving further up the gradient to health care, there exists a growing concern that a continued distancing of humans from exposure and interactions with nature, or nature-deficit disorder (Louv, 2011), is leading to an array of health issues. The ability of nature to heal and foster well-being has been advanced by work investigating "therapeutic landscapes" (Gesler, 2005, 2012; Williams, 1999), or "the use of particular places for the maintenance of health and well-being" (Velarde, Fry, & Tveit, 2007, p. 200). Of particular note is the potential of natural landscapes to improve healing in health care facilities. The greening of hospitals and elder care facilities can provide salutogenic, or health enhancing, benefits. A stunning example of this is Khoo Teck Puat hospital in Singapore (Figure 1.7). This "hospital in a garden" and "garden in a hospital" has incorporated nature in its design to enhance

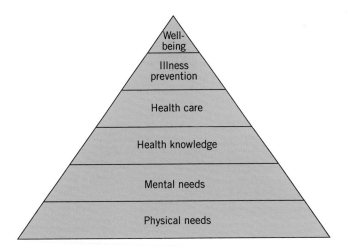

Figure 1.6: The health gradient

Source: Adapted from Laughlin and Black (1995).

healing in patients and also the larger environment, acknowledging that the two are inseparable (Khoo Teck Puat Hospital, 2015). Our knowledge of the ability of the natural environment to promote healing has developed to such a degree that investigations into the health benefits of "forest bathing," or accessing GI, in Japan have revealed that the atmospheric composition of forests is linked to physiological responses in immune system functioning including the production of anti-cancer proteins and the reduction of blood glucose levels in diabetic patients (Li et al., 2008a, 2008b; Ohtsuka, Yabunaka, & Takayama, 1998).

Moving another step up the pyramid, interactions with GI may also result in the prevention of illness through GI's ability to support behaviors such as physical activity. Green infrastructure is also associated with stress reduction, and exposure to GI is therefore a preventive measure of reducing the risks of the many diseases exacerbated by regular and sustained stress. The therapeutic and preventive potential of GI culminates in the well-being necessary to achieve health as the World Health Organization (WHO) defines it. Green infrastructure supports the ecosystems services that permeate all levels of the health gradient. The value of

Figure 1.7: Khoo Teck Puat Hospital, Singapore
Source: Jui-Yong Sim.

GI in supporting the various ecosystem services can be captured in quantifiable reductions in disease, and this can be monetized. Monetizing the often overlooked health value of ecosystem services adds weight to the benefits side of the equation when weighing the costs and benefits of GI conservation.

The valuation of the health benefits of ecosystem services

Ecosystem service valuation is the science, coupled with a heavy dose of art, of attributing monetary values to the services and products ecosystems provide to humans (benefits) or the cost to humans when these services are compromised. Ecosystem service valuation is briefly touched on here for two reasons. First, we will see in subsequent chapters that value estimates have been placed on GI and ecosystem services as they relate to water management, food production, and carbon capture. Some estimates have also been extended to air quality and energy consumption issues as well as considerations of "livability" (Gallet & Grant, 2010). Second, monetary values are being placed on GI and ecosystem services, but the potentially enormous costs to human health when ecosystem services are compromised is often ignored in these calculations. The exclusion of health represents a narrow view of ecological complexity and an underappreciation for the multitasking of nature and ecosystems (Wegner & Pascual, 2011; Wolf, 2014). The omission of health also reveals the apparent disconnect between the natural environment and human health and well-being; a disconnect apparent in the very method by which the value of ecosystem services to human functions and life are estimated. The calculated value of GI would undoubtedly be much greater if health was consistently considered as a benefit of GI. For example, it has been found that 24 percent of the global burden of disease could potentially be prevented with environmental improvements to air and water quality (Prüss-Ustün, Bonjour, & Corvalán, 2008). This is a startlingly large proportion considering that it only reflects environmental improvements in two key areas. What is excluded from such calculations are "environmental improvements" that conserve the GI that supports air and water quality. If we were to add to air and water the myriad other health-promoting ecosystem services GI supports (e.g. food, mental health), the preventable global burden of disease attributed to the environmental improvement of GI conservation would be much, much higher. This would likely remain true even with taking into account the small proportion of the world's population that can afford to implement the measures which mediate environmental threats.

Measures such as the Inclusive Wealth Index (IWI) capture the relationship between changes in natural capital and manufactured and human capital (Duraiappah, 2014). Health is included in the human capital component of the IWI, but only with the crude measure of longevity. When more inclusive measures of health capital are included, they are "exceedingly dominant" (p. 31) and are therefore omitted. Methods for including measures of health beyond longevity in future IWI calculations are underway. The IWI holds promise for providing a more accurate assessment of the value of health to human capital and its relation to natural capital and ecosystem services.

In instances when a narrow view of ecosystem complexity has been overcome and human health has been recognized as a valuable benefit of ecosystem services, it is at times still omitted from final cost–benefit figures due to the noted difficulty in linking disruptions in ecosystem services to health outcomes (van Essen et al., 2011, p. 69). Because it is difficult to measure, we exclude one of the most important and costly outcomes to society: its health. Calculating values for ecosystem services is a major feat fraught with an uncertainty not hidden from the purview of the evolving science of environmental economics, but the danger of omitting health from ecosystem service cost–benefit analyses is that the value of the GI that supports these services is then greatly underrepresented. This impacts people's lives because weighing monetary costs and benefits influences how policy is formulated and decisions are made. When GI is undervalued and compromised in favor of "more valuable" resources, ecosystem services and health are compromised. A moral argument could certainly be made against valuing the natural environment and human health in monetary terms, but not including the health benefits of ecosystem services in cost–benefit analyses on moral grounds only allows human health to continue to be ignored. While we await a moral awakening, it is likely a more productive course of action to try to improve cost–benefit analyses with a more complete approach that includes health.

There are interactive tools such as BenMAP, focused on providing the health and economic costs and benefits with changes in air quality, that are on the leading edge of remedying the common exclusion of health in environmental valuation (www.epa.gov/air/benmap/).[11]

Released in 2006 by the US Department of Agriculture (USDA) Forest Service:

> i-Tree is a state-of-the-art, peer-reviewed software suite from the USDA Forest Service that provides urban forestry analysis and benefits assessment tools. The i-Tree Tools help communities of all sizes to strengthen their urban forest management and advocacy efforts by quantifying the structure of community trees and the environmental services that trees provide.

The services trees provide are categorized in storm water, carbon sequestration, and air quality. In the i-Tree calculation of air quality services, there is some accounting of the health benefits of nitrogen dioxide (NO_2), sulfur dioxide (SO_2), and volatile organic compounds (VOCs) reduction. These estimates are based on BenMAP and IMPACT (van Essen et al., 2011) (www.itreetools.org/).

When health *is* included in cost–benefit calculations of GI, it has been found to be highly significant. A cost–benefit analysis of walking and cycling infrastructure that includes greenways calculated the cost savings of avoiding long-term ailments through physical activity (Sælensminde, 2004). Even very conservative estimates show that the benefits far outweigh the costs of implementing walking and cycling infrastructure projects with the single greatest benefit being to health. Advancing these types of much needed studies could involve adding the ecosystem service GI provides in capturing air pollution and supporting the walking and biking behaviors that mitigate the release of air pollution, both of which would undoubtedly add to the health value of GI.

To fully represent the interdependencies of all these services in an ecological fashion, ecosystem service valuation models would also need to capture the influence diminished GI has on a more expansive set of ecosystem services and health benefits. For example, the monetized health benefits of GI to air quality would need to capture not only the cost of treating asthma but also the other conditions associated with poor air quality such as nutritional deficiencies that arise from reduced crop yields, compromised water quality, rising temperatures, decreased levels of physical activity due to pollution making outdoor recreation unattractive, and the mental health and social cohesion benefits that are lost as a result of the diminished attractiveness of GI. This is a much more accurate reflection of the costs to health, and more accurate models that reflect these cumulative values are necessary if informed decisions are going to be made about the vast array of overlapping and interdependent human health benefits of conserving GI.

The common exclusion of health from ecosystem service valuation mirrors the often overlooked importance of the natural environment in the ecological model of health. As we will see in Chapter 2, the evolving ecological models of health recognize (in theory) the fundamental importance of the natural environment in creating the conditions for humans to be healthy. Let us first look at a number of studies that have demonstrated relationships between GI and improved health, all of which could be broadly classified as biophilic epidemiology.

Biophilic epidemiology

The biophilia hypothesis (Kellert & Wilson, 1993; Wilson, 1984) posits an innate biological connection between humans and elements of nature and the inclination of humans "to affiliate with natural systems and processes instrumental in their health and productivity" (Kellert, Heerwagen, & Mador, 2008, p. vii). *Biophilic epidemiology* is then the study of how the human affiliation with nature affects health and well-being. This is an expansion on the typical approach to epidemiology which is the study of the distribution and determinants of disease. Biophilic epidemiology considers the causes of health and well-being and not just the causes of disease. What is slowly being revealed in studies that could be classified under the umbrella of biophilic epidemiology is that diminished GI and isolation from the natural environment exacts a price on health. Environments that inhibit the ability to affiliate with nature, and actions that compromise the ability of GI to continue to provide essential ecosystem services, result in diminished health and well-being.

What constitutes a viable natural habitat to support health today has not evolved far beyond what our distant ancestors relied on to sustain their health and survival. The habitat our ancestors inhabited was exclusively, and until only recently (evolutionary speaking), dominated by the natural environment. Ignoring the importance of the natural environment in human health and well-being is ignoring eons of the human affiliation with nature and the likelihood that the need for this affiliation is still present in contemporary human beings. There is something inherently and universally appealing about nature, but we cannot attribute this affinity solely to genetic predisposition. "Biophilia … is probably not an attribute with a strong penetrance," or frequency of genetic expression (Grinde & Patil, 2009, p. 2338). Biophilic tendencies are likely attributable to sociocultural factors as well and not solely evolutionary factors fixed and static over the course of human evolution (Hartig, 1993).

The human desire to affiliate with the natural environment is witnessed with people voting with their feet and their wallets. Living near publically accessible GI often holds a high real estate premium, and people consistently vote to spend large sums to conserve GI in their communities (Land Trust Alliance, 2012). Even in countries where technology and infrastructure can create a sense of separation from nature, perceptions of the role GI plays in people's lives is overwhelmingly positive (Croucher, Myers, & Bretherton, 2007). As representations of nature disappear from one's living environment, people are more likely to leave artifact-filled environments on holiday to seek out nature (Sijtsma, de Vries, Van Hinsberg, & Diederiks, 2012). A biophilic approach puts forth that there may be a tacit motive for self-improvement and self-preservation underlying the demonstrated desire to affiliate with and protect the natural environment that spawned us.

> The connection to nature is evident in sociocultural activities that humans still feel compelled to do even though they could only seldom be considered functional. We "cross the ocean in artificially primitive boats, climb mountains we could fly over, and kill animals we do not eat" (Lindheim & Syme, 1983, p. 350; Slater, 1970).

This is not to imply that any complete move "back to nature" will guarantee health and safety. The billions of humans worldwide who live amidst abundant nature could attest that nature can be fickle, cruel, and deadly. As we are an organism of nature, so too are we susceptible, either through misfortune or ignorance, to the elements of nature that can harm us. Indeed, the complement to biophilia, biophobia, recognizes the innate response to natural elements and environments that are threatening (Ulrich, 1993). Keeping with the theme of this book, I would propose that the perceived threat is that to health and welfare. The findings of what could be termed *biophilic epidemiology* research continue to reveal how the natural environment supports health, but these findings do not imply that we should cease to protect ourselves from threats because they are natural; we should continue to wash naturally occurring bacteria from our hands before eating. What biophilic epidemiology research reveals is that when humans are denied the ability to affiliate with nature, and when the natural environment's ability to support ecosystem services is degraded, human health suffers. To sustain our great strides in increasing longevity, our reliance on the natural environment cannot be ignored, but measures to satisfy biophilic needs and improve health with GI must not be clouded by a utopian vision of pristine and friendly nature. The gray infrastructure that can degrade the natural environment can also protect us from its threats to health.

There are nine categories in a typology of biophilic values (utilitarian, naturalistic, ecologistic–scientific, aesthetic, symbolic, humanistic, moralistic, dominionistic, negativistic) (Kellert & Wilson, 1993). Among these, it is the utilitarian and naturalistic that hold the greatest potential for understanding the relationship between biophilic tendencies and health. The *utilitarian* refers to the exploitation of nature. The most obvious examples of this are the food, water, and air that no human can survive without, but it also includes an ever expanding list of ecosystem services that are threatened when the quality of the natural environment is diminished. The *naturalistic* refers to the satisfaction achieved from exposure to nature. It is the naturalistic that captures the mental health benefits derived from contact with nature and the physical fitness gained from accessing it.

Although it is the utilitarian and naturalistic that are supported by current evidence as having an impact on health, it is the "cumulative, interactive, and synergistic impact" of all these values that "contribute to the possibility of a more fulfilling personal existence" (Kellert & Wilson, 1993, p. 60). If it is the affiliation with nature that leads to fulfillment, and it is fulfillment that contributes to well-being, then it is the natural environment that helps us achieve health as defined as a complete state of well-being. Although it is the utilitarian and naturalistic that have been studied for their relation to health, it is likely a more inclusive synergy of biophilic values that is evidenced in the stated ability of GI to improve quality of life and life satisfaction (CABE Space, 2005; White, Alcock, Wheeler, & Depledge, 2013). We see that people with a greater acceptance of biophilic values and connection to nature, or "nature-relatedness," evidenced in activism and behavior to protect it, have improved overall subjective well-being, most notably sustained vitality (Nisbet, Zelenski, & Murphy, 2011). It is outcomes such as this that conform to the WHO definition of health as well-being and that are in great need of further biophilic epidemiology study.

Research testing the biophilia hypothesis more broadly, not biophilic epidemiology, has taken many forms encompassing the examination of nature as the context for "human maturation, functional development, and ultimately survival" (Kellert et al., 2008, p. vi). Most biophilia research that I am aware of examines people's preference for and level of functional development in natural environments compared to environments where nature is diminished or absent. The focus here on biophilic epidemiology takes the logical next step, working under the hypothesis that when the innate human preference for affiliation with the natural environment is satisfied, human health is improved. Affiliation is achieved not only through access and exposure to GI. Affiliation also extends to recognizing the unity between health and the mere presence of GI that may never be accessed or viewed. The ability to affiliate with nature by accepting a dependence upon regional and global GI systems is fundamental to their protection and ability to continue to support health.

The extant literature measuring health and well-being as outcomes of one's affiliation with GI is growing. The methods employed in this type of research examining the biophilic conditions, activities, attitudes, and institutions necessary for health (Beatley, 2011, see Box 3.1) have steadily increased in their sophistication. For example, a recent study employed biomarkers to determine the stress reduction properties of exposure to nature (Ward Thompson et al., 2012). Despite this, much work needs to be done, and guidance has been provided on the kind of research needed to advance the often lauded, but grossly understudied, health benefits of the natural environment and processes (Beatley, 2008). Let us now look at some of the extant biophilic epidemiology research that focuses on overall health and well-being as outcomes before delving into subsequent chapters that address the specific ecosystem services of GI and how they support health.

Before doing this, it is important to note four things. First, it is worth reminding ourselves again that health encompasses physical, mental, and social well-being that are independently significant and interdependent in determining health. Outcomes such as perceived health, quality of life, and possibly longevity are likely better indicators of health as the WHO defines it. Second, the GI and health literature has been dominated by studies examining the role of GI in creating the conditions (e.g. for mental restoration) and supporting the behaviors (e.g. physical activity) associated with reducing chronic health issues or their precursors. There is also quite a bit of literature to support the role of GI in infectious disease ecology, but this is almost always overlooked in overviews of the pathways that connect GI and health. Third, different effects on health are realized over very different time frames. There are discrete and continuous interactions with GI that result in immediate benefits to health and other benefits that accrue over the lifecourse (Astell-Burt, Mitchell, & Hartig, 2014). Walking once a year in a lush park may provide immediate benefits in mental restoration, but this infrequent exposure will do little in protecting against conditions that build over the lifecourse (e.g. heart disease). Long-term benefits require regular access to local GI. Regional GI systems are continuously functioning to cycle water and cleanse the air, and there are both immediate and long-term benefits to these ecosystem services. Fourth, the benefits of GI change over the lifecourse as contexts for interaction with GI change. As a child, a backyard, neighborhood park, vacant lot, or schoolyard may be the context for access and exposure. As a teenager and young adult, it may be a city street. A young couple may gain exposure to GI on the commute to work and access to GI in a local park, and their children from the aforementioned contexts. As kids leave the house and careers come to an end, the housing choices of older adults will affect whether the benefits accrued over the lifecourse will continue into the years when they could indeed be the most important.

Two sources of guidance on how to satisfy the biophilic needs of urban dwellers are: *Biophilic Design: The Theory, Science and Practice of Bringing Buildings to Life* (Kellert et al., 2008) and *Building for Life: Designing and Understanding the Human–Nature Connection* (Kellert, 2005).

A group in the Netherlands has done quite a bit of commendably rigorous work studying the health benefits of GI (Groenewegen, van den Berg, de Vries, & Verheij, 2006), and how a lack of "vitamin G" (G being Green) influences health (Maas, Verheij, Groenewegen, de Vries, & Spreeuwenberg, 2006). They have found that people who live in greener environments have better perceived health than persons in less green environments (Figure 1.8). This is especially true for segments of the population that are more likely to spend a greater proportion of their time close to home and where the residential environment is therefore the dominant context of daily life: the young, the old, women who work in the home, and those of lower socioeconomic status (de Vries, Verheij, Groenewegen, & Spreeuwenberg, 2003; Maas et al., 2006). Additionally, in Denmark, the distance to greenspace and its availability in one's local environment seems to make difference to health. People living further than 1km away from the nearest greenspace had poorer reported health and health-related quality of life than respondents living closer than 1km from greenspaces, implying access and exposure are important (Stigsdotter et al., 2010). The use of surveys of perceived health status should not be viewed as a limitation. Health and well-being are very subjective states. Although measures of perceived health may not always be valid to measure the absence of disease, the definition of health reminds us that this is not solely what we are striving for. To appease those who rightfully suggest that objective measures

Figure 1.8: Monseigneur Nolenspark, Maastricht, the Netherlands
Source: Jorge Franganillo.

are useful complements to subjective measures, the health benefits of greenspace have also been associated with reduced physician-assessed morbidity (Maas et al., 2009). Again, this is not a measure of health as much as a measure of the presence or absence of clinically diagnosed disease. The conclusion from the years of work done by the group in the Netherlands is that GI improves health through the mechanisms of reducing stress and increasing social cohesion (Groenewegen, van den Berg, Maas, Verheij, & de Vries, 2012), two mechanisms that certainly have an effect on the physical, mental, social, and spiritual well-being public health has made its mission to strive for.

A handful of studies have examined the relationship between GI and mortality. These studies aim to examine the cumulative effects that disease and threats to well-being may have on longevity. One such study examined the relationship between a number of measures of GI and all-cause and cardiovascular mortality in Florida (Coutts, Horner, & Chapin, 2010). While an association was found between having greater amounts of GI within defined distances from home and reduced all-cause and cardiovascular mortality, there was not a significant relationship found between the gross amount of greenspace (in a county) nor distance to the *nearest* greenspace and either measure of mortality. The inherent limitation of the use of large spatial units to represent "home" (census tracts) was addressed in a follow-up study which examined individual death records and measured the distance to and amount of greenspace from an individual's residence (Coutts & Horner, 2015). In this study, increased distance to the nearest greenspace was found to be a significant determinant of premature mortality. A study done in the UK using a percent area coverage of greenspace (in spatial units similar to census tracts in the USA) coincides with the finer scale individual level results (Mitchell & Popham, 2008). The UK study also demonstrated the equalizing effect of greenspace in reducing health inequalities in all-cause and circulatory mortality. Comparing the complementary results of the UK and Florida studies raises the important issue of scale. While spatial units of analysis beyond the individual are worthy of exploratory analyses and make an important contribution to the guidance of policies usually applied to larger geographic units, two mechanisms by which GI operates to support health (i.e. access and exposure) are more accurately represented at the individual level. A case in point is one study conducted at the city scale in the USA that found that all-cause mortality was actually higher in greener cities (Richardson et al., 2012). The authors note that the sprawling nature of US cities may make them more green, but the behaviors that are associated with this style of built environment (e.g. auto dependence) may supersede the health benefits of greenness. What is lost at the city scale or other large units of analysis is the distribution of GI that influences everyday exposure and access. Some neighborhoods within a city may have no GI while in others there may be ample GI, but these neighborhood differences are lost when aggregating up to larger spatial units. What is also lost in aggregation is the type and quality of GI in neighborhoods. This is important because in neighborhoods with poorer health status, parks have been found to be of much poorer quality (Coen & Ross, 2006). Green infrastructure may be present, but its condition and design may inhibit the health benefits derived from its use. Noting the type of GI and its distribution at finer scales of analysis (e.g. individual, neighborhood) is necessary to accurately capture factors that influence the health benefits of access and exposure.

Two studies done in Japan, a very different cultural environment than that of the previously cited studies, also examine the importance of GI on mortality. In Tokyo, older persons with walkable greenspace in their neighborhood experienced increased longevity (Takano, Nakamura, & Watanabe, 2002). Across all Japanese cities, where almost three-quarters of the population lives, a lack of vegetation was associated with an increase in female, but not male, mortality rates (Fukuda, Nakamura, & Takano, 2004). This coincides with the findings from the group in the Netherlands that found a particular positive benefit for females. Again, what is lost when we scale up the level of analysis, even though Fukuda et al. (2004) disaggregated cities into administrative wards, is the distribution and type of GI in one's local, everyday environment. The Takano et al. (2002) study suggests that if GI is easily accessible, at least by older persons, it is important for increasing longevity in both males and females.

Finally, representing the other end of the lifecourse spectrum, there is also evidence from Spain that suggests GI close to home supports better pregnancy outcomes (Dadvand & Sunyer, 2012). So, taken together with the above studies that measure overall health and mortality, having the opportunity to affiliate with the natural environment positively influences health at various stages of the lifecourse: birth, life, and delaying death. The results from all of the above studies are especially significant when GI is part of an individual's everyday environment and the opportunity for access and exposure is therefore increased. Larger scale analyses are still very useful in capturing the benefits derived from the presence of GI at scales beyond one's everyday environment. It is the presence of GI at these scales that is necessary to support many basic ecosystem services often overlooked in studies of GI and health.

Green infrastructure is a clear case of returning to the foundation of what we need to survive and thrive as a species. In short, GI is absolutely necessary to ensure health. The acknowledgment of the permeating and complex ways that GI supports health has necessitated the bridging of disciplinary barriers and the birth of a number of interdisciplinary movements.

Movements to advance a greener ecological model of health

This is a pivotal and exciting time for those working at the intersection of the natural environment and health. There are a number of movements (Table 1.3) that are transcending disciplinary silos in order to unravel the ecological complexity of the relationship between the natural environment (including GI), ecosystems, and health. The ecological models of health presented in Chapter 2 capture the influence of the natural environment on health, but the movements listed here extend the often uni-directional focus of these models to include the reciprocity, and therefore more accurately the ecology, of this relationship. These communities of research, practice, and thought could be considered appropriate "homes" for the study of the natural environment and health. They are composed of the "wide variety of actors," including those not trained to treat symptoms of ill-health, absolutely necessary to more fully understand a truer ecology of human health (Hartig, Mitchell, de Vries, & Frumkin, 2014, p. 21.3; Van Herzele, Bell, Hartig, Podesta, & van Zon, 2011).

What may seem a glaring omission in Table 1.3 is the public health specialization of Environmental Health (EH). This is not a mistake. The field charged with protecting the public from environmental threats has been somewhat slow in adopting an ecological perspective that includes the natural environment. With calls for the field to move in this direction (Frumkin, 2001; Parkes, Panelli, & Weinstein, 2003), and the public served by EH attributing GI (e.g. urban greenways) to improved health (e.g. Lindsey & Knaap, 1999), progress has been made in moving towards a greater consideration of landscapes and ecosystems in EH. That said, the toxicological approach to environmental health that focuses on how elements in the environment can harm health still reigns (Coutts & Taylor, 2011; Frumkin, 2001; Myers et al., 2013). Much less attention is given to how the natural environment can support health. This has been criticized as similar to

Table 1.3: Summary of movements that connect health and the natural environment

Movement	Description	Reference
Conservation Medicine	Connects ecosystem, animal, and human health, but largely focused on infectious diseases that are exacerbated by human manipulation of animal habitat	Aguirre, Ostfeld, & Daszak (2012)
EcoHealth	Mission: to strive for sustainable health of people, wildlife, and ecosystems by promoting discovery, understanding, and transdisciplinarity	Wilcox et al. (2004)
OneHealth	Recognizes that the health of humans is connected to the health of animals and the environment. Like conservation medicine, focused on infectious disease	American Veterinary Medical Association (2008)
Human Ecology	Study of human–environment interactions that has been extended to health outcomes	Parkes, Panelli, & Weinstein (2003)
Health Ecology	Health ecology situates health as the core concept in human ecology, the ultimate aim of which "is the creation and maintenance of healthy people in healthy environments"	Honari & Boleyn (1999, p. 17)
Ecotoxicology	This is important for understanding the ecological products and mechanisms that make ecosystems and humans sick	DiGiulio & Monosson (1996)
Public Health Ecology	Landscape structure influences the agents and stressors that have direct and indirect influences on health	Coutts (2010)

Source: Author.

"looking at all the side effects of a drug without understanding its efficacy" (Bird, 2007, p. 113). Tremendous advances have been made in understanding how air and water pollutants adversely affect health (often, how much we can tolerate without getting sick). Far less attention has been given to role of the natural environment in removing pollutants or mitigating their release. For example, we know how some pesticides adversely affect health, but much less about how the GI that supports biodiversity affects the behavior of pests. Infectious disease ecology (discussed in Chapter 5) is one area where the landscape has been considered as an upstream determinant of environmental threats to health. Environmental health specialists are partners in many of the movements listed in Table 1.3, and their continued involvement in these movements makes it more likely that nature-based health promotion (Hansen-Ketchum & Halpenny, 2011) will gain more prominence in EH. As it does, EH has the potential to become an umbrella for human–environment interactions focused on health.

In the meantime, a number of international collaborations have been formed to study and develop policies to protect the natural environment for the purpose of protecting human health. The Co-Operation On Health And Biodiversity (COHAB) initiative, established in 2005 with its Secretariat based in Ireland, is one such international organization. In no uncertain terms, they state that

> human health is one of the most important indicators of sustainable development—a healthy human population is dependent upon a healthy natural environment. The ecosystem approach to health recognizes the intimate links between ecosystem services, human and animal health, biological diversity, and social and economic development, and recognizes the need for integrated public health policies and development programs that view the protection of ecosystems as an important part of achieving their objectives.
>
> (COHAB, n.d.)

Other communities of researchers examining human–environment interactions may also be appropriate "homes" for the study of the natural environment and health. Coupled human and natural systems (CHANS), not listed in Table 1.3 because it is not focused specifically on human health, has received a swelling of interest over the past decade (Liu, Dietz, Carpenter, & Folke, 2007; Liu et al., 2007). Coupled human and natural systems is an umbrella to the human–environment systems, ecological–economic systems, population–environment systems, or ecological research that overlap one another in various complex ways. Distinguishing the CHANS approach "is an explicit acknowledgement that human and natural systems are coupled via reciprocal interactions" (McConnell et al., 2011, p. 219). Again, not focused on health outcomes, a CHANS approach recognizes that the reciprocity between human and natural systems affect the "potable water, clean air, nutritious food, raw materials, and medicine" on which the human at the center of ecological models of health depends (Liu, Dietz, Carpenter, & Folke, 2007, p. 639). It is exactly this reciprocity that is described in the Millennium Ecosystem Assessment *Ecosystems and Human Well-Being: Health Synthesis* (2005a). Recognizing that these interactions cannot be understood within the confines of any single discipline, it is communities of researchers like CHANS that are moving towards the consilience of knowledge necessary to deliver health. Due to the complexity of these relationships, GI and health is an area of necessary collaboration between disciplines including public health, geography, forestry, atmospheric sciences, ecology, urban planning, public policy, and many others. This consilience of disciplines is essential to understanding the ecosystem components and processes and the services they deliver and to manage the human actions that threaten them. This is not to say that expertise in any single discipline is not necessary and useful. Rather, by bridging disciplinary expertise, it is much more likely that overlapping goals and needs can be identified. Depending on the discipline, the focus of human–environment interactions may be environmental quality and viable ecosystems or human outcomes, but being part of a community forces one to continually consider how they are inextricably linked.

The consilience fostered by these communities is most impactful on people's lives and health if the questions being addressed are inspired by their potential application. This "use-inspired basic research" would aim to "probe unknown fundamentals and meet a societal need" (Clark, 2007; Stokes, 1997, p. 83). This focus on application is distinct from basic research which aims to advance knowledge void of a consideration of its application. It is also distinct from applied research which aims to consider application void of any consideration of theory. The unknown fundamentals in the case of GI and health are the various mechanisms by which humans influence and are influenced by GI, and the societal need is environmental protection that creates the conditions necessary for health. A use-inspired approach that considers use and policy first would increase the value to society (Myers et al., 2013, p. 18759). Ideally, with increasing amounts of use-inspired evidence demonstrating the fundamental importance of GI to health, further action to conserve GI will be taken with benefits to humans and the natural environment. This is not to imply that irrefutable proof is necessary before any action is taken. A precautionary approach challenges us to weigh the cost to health of not taking action against our increasingly nuanced understanding of GI's sweeping benefits.

Summary

A GI system is composed of landscape patches and corridors that together form a landscape matrix. While all individual components of GI provide some health benefits, an interconnected system of patches and corridors creates the landscape structure that is more adept at delivering the ecosystem services on which health and well-being depend. Landscape function and ecosystems depend on landscape structure. The definition of GI adopted here includes landscape components but also other representations of nature such as those introduced into urban environments (e.g. street trees, vegetated medians). These representations of nature also contribute to the GI system and ecosystem services with valuable health benefits.

Health is commonly overlooked in the valuation of ecosystem services. This is not due to ignorance, but rather because health is difficult to isolate and measure, and because when health is included in monetary valuations of ecosystem services—even crudely represented with one or two potential risk factors—it dominates and clouds other evaluated factors. Because ecosystem services are so valuable to health, we tend to ignore this relationship. The sweeping benefits of GI and ecosystem services to health are increasingly revealed in what could be coined biophilic epidemiology, or the study of the relationship between the innate human preference and dependence on GI and how its presence, and access and exposure to it, influence health. A number of inter- and trans-disciplinary movements, composed of a wide variety of actors, have been formed to study the complexity of this relationship.

Notes

1. I was unable to find the original document containing this quote. Any lead to the original source of this quote would be most welcome.
2. Not included in this book, but certainly important to a more complete accounting of the components of the natural environment essential to health, are water features or blue infrastructure. Water is discussed in Chapter 3, but only in how green infrastructure is essential in supporting its quality and abundance. The benefits of marine and coastal ecosystems are presented in *Marine and Coastal Ecosystems and Human Well-Being: A Synthesis Report Based on the Findings of the Millennium Ecosystem Assessment* (UNEP, 2006).

3. Important to differentiate here is green infrastructure, as defined in this book, and the greening of gray infrastructure. Gray infrastructure could be termed "green" if it increases efficiencies and reduces its impact on the natural environment. This is *not* how green infrastructure is used here.

4. For a more in-depth description of these principles, I recommend *Land Mosaics: The Ecology of Landscapes and Regions* (Forman, 1995).

5. Pretty et al. (2005) have summarized these three levels of engagement slightly differently into viewing, presence, and active participation.

6. Rouse and Bunster-Ossa (2013, pp. 42–3) also provide a table of various typologies of green infrastructure at various scales but in a matrix that shows the relationship between green infrastructure and their six planning and design principles.

7. A landscape ecology approach categorizes the services of green infrastructure slightly differently into Abiotic, Biotic, and Cultural (ABC) services (Ahern, 1995, 2007). *Abiotic* services, such as those provided by a network of green streets, include surface and groundwater interactions and improvement of hydrological regimes, carbon and greenhouse gas sequestration and storage, and energy conservation through the buffering of climatic extremes. *Biotic* services comprise habitat enhancement, support of flora and fauna interactions, and biomass production. *Cultural* services involve support of shaded multimodal circulation (especially walking and bicycling), safer walking, community character, and enhanced opportunities for healthy social interactions. The approach adopted by the World Health Organization and other sources captures these services in a slightly different fashion.

8. WHO notes the aesthetic benefits of nature under the cultural domain (Millennium Ecosystem Assessment, 2005a, p. 14, Fig. 1.3). Kellert and Wilson (1993) differentiate the *naturalistic* and *aesthetic* in their nine dimensions of biophilic tendencies, situating mental benefits in the naturalistic. I see no benefit in belaboring this point other than simply using it as an example of the array of categorizations that exist. These differentiations are likely important for parsing theory but not for gaining a general understanding of these principles.

9. An example of how these services are divided into categories for urban regions is provided in Niemelä et al. (2010, Table 1).

10. The habitat or supporting service of biodiversity is not presented in a stand-alone chapter. It appears most notably in the section on medicine in Chapter 3 and in Chapter 5 on infectious disease.

11. BenMAP uses the van Essen et al. (2011) estimates to measure the health impacts from volatile organic compounds (VOCs).

References

Aguirre, A. A., Ostfeld, R., & Daszak, P. (2012). *New directions in conservation medicine: Applied cases of ecological health.* New York, NY: Oxford University Press.

Ahern, J. (1995). Greenways as a planning strategy. *Landscape and Urban Planning, 33*(1–3), 131–55.

Ahern, J. (2007). Green infrastructure for cities: The spatial dimension. In V. Novotny & P. Brown (Eds.), *Cities of the future: Towards integrated sustainable water and landscape management* (pp. 267–83). London: IWA Publishing.

American Veterinary Medical Association. (2008). *One health: A new professional imperative. One Health Initiative Task Force: Final Report.* Retrieved December 8th, 2016 from https://www.avma.org/KB/Resources/Reports/Pages/One-Health.aspx

Astell-Burt, T., Mitchell, R., & Hartig, T. (2014). The association between green space and mental health varies across the lifecourse. A longitudinal study. *Journal of Epidemiology and Community Health, 68,* 578–83.

Barton, H., & Tsourou, C. (2000). *Healthy urban planning.* London: Spon Press.

Beatley, T. (2008). Toward biophilic cities: Strategies for integrating nature into urban design. In S. R. Kellert, J. Heerwagen, & M. Mador (Eds.), *Biophilic design: The theory, science, and practice of bringing buildings to life* (pp. 277–95). London: John Wiley.

Beatley, T. (2011). *Biophilic cities: Integrating nature into urban design and planning.* Washington, DC: Island Press.

Benedict, M. A., & McMahon, E. T. (2002). Green infrastructure: Smart conservation for the 21st century. *Renewable Resources Journal, 20*(3), 12–17.

Benedict, M. A., & McMahon, E. T. (2006). *Green infrastructure: Linking landscapes and communities.* Washington, DC: Island Press.

Bird, W. (2007). *Natural thinking: Investigating the links between the natural environment, biodiversity and mental health.* Sandy, UK: Royal Society for the Protection of Birds.

CABE Space. (2005). *Decent parks? Decent behaviour? The link between the quality of parks and user behaviour.* London: Commission for Architecture and the Built Environment.

Chivian, E., & Bernstein, A. (2008). *Sustaining life: How human health depends on biodiversity.* New York, NY: Oxford University Press.

Clark, W. C. (2007). Sustainability science: A room of its own. *Proceedings of the National Academy of Sciences, 104*(6), 1737–8.

Coen, S. E., & Ross, N. A. (2006). Exploring the material basis for health: Characteristics of parks in Montreal neighborhoods with contrasting health outcomes. *Health & Place, 12*(4), 361–71.

COHAB. (n.d.). Background. Retrieved May 19, 2015, from www.cohabnet.org/en_about_background.htm

Coutts, C. (2010). Public health ecology. *Journal of Environmental Health, 72*(6), 53–5.

Coutts, C., & Taylor, C. (2011). Putting the capital "E" environment into ecological models of health. *Journal of Environmental Health, 74*(4), 26–9.

Coutts, C., & Horner, M. (2015). Nature and death: An individual level analysis of the relationship between biophilic environments and mortality in Florida. In P. Kanaroglou, E. Delmelle, & A. Paez (Eds.), *Spatial analysis and health geography* (pp. 295–312). Farnham: Ashgate.

Coutts, C., Horner, M., & Chapin, T. (2010). Using geographical information system to model the effects of green space accessibility on mortality in Florida. *Geocarto International, 25*(6), 471–84.

Croucher, K., Myers, L., & Bretherton, J. (2007). *The links between greenspace and health: A critical literature review.* Stirling: Greenspace Scotland.

Dadvand, P., & Sunyer, J. (2012). Surrounding greenness and pregnancy outcomes in four Spanish birth cohorts. *Environmental Health Perspectives, 120*(10), 1481–7.

de Groot, R. S., Wilson, M. A., & Boumans, R. M. (2002). A typology for the classification, description and valuation of ecosystem functions, goods and services. *Ecological Economics, 41*(3), 393–408.

de Groot, R. S., Alkemade, R., Braat, L., Hein, L., & Willemen, L. (2010). Challenges in integrating the concept of ecosystem services and values in landscape planning, management and decision making. *Ecological Complexity, 7*(3), 260–72.

de Groot, R. S., Fisher, B., & Christie, M. (2010). Integrating the ecological and economic dimensions in biodiversity and ecosystem service valuation. In P. Kumar (Ed.), *The economics of ecosystems and biodiversity: Ecological and economic foundations* (pp. 1–40). London: Earthscan.

de Vries, S., Verheij, R. A., Groenewegen, P. P., & Spreeuwenberg, P. (2003). Natural environments—healthy environments? An exploratory analysis of the relationship between greenspace and health. *Environment and Planning A, 35*(10), 1717–32.

DiGiulio, R. T., & Monosson, E. (Eds.). (1996). *Interconnections between human and ecosystem health.* London: Chapman & Hall.

Dramstad, W. E., Olson, J. D., & Forman, R. T. T. (1996). *Landscape ecology principles in landscape architecture and land-use planning.* Washington, DC: Island Press.

Duraiappah, A. (2014). Inclusive wealth: Incorporation of health information. In S. Landi (Ed.), *Including health in global frameworks for development, wealth, and climate change: Workshop summary* (pp. 30–3). Washington, DC: The National Academies Press.

Ecosystem. (2013). Retrieved from http://oxforddictionaries.com/us/definition/american_english/ecosystem

Forman, R. T. T. (1995). *Land mosaics: The ecology of landscapes and regions.* New York, NY: Cambridge University Press.

Frumkin, H. (2001). Beyond toxicity: Human health and the natural environment. *American Journal of Preventive Medicine, 20*(3), 234–40.

Fukuda, Y., Nakamura, K., & Takano, T. (2004). Wide range of socioeconomic factors associated with mortality among cities in Japan. *Health Promotion International, 19*(2), 177–87.

Gallet, D., & Grant, J. (Eds.). (2010). *The value of green infrastructure: A guide to recognizing its economic, environmental and social benefits.* Chicago, IL: Center for Neighborhood Technology.

Gesler, W. (2005). Therapeutic landscapes: An evolving theme. *Health and Place, 11*(4), 295–7.

Gesler, W. (2012). Therapeutic landscapes: Medical issues in light of the new cultural geography. *Journal of Personality and Social Psychology, 34*(7), 735–46.

Girling, C., & Kellett, R. (2005). *Skinny streets and green neighborhoods: Design for environment and community.* Washington, DC: Island Press.

Grinde, B., & Patil, G. G. (2009). Biophila: Does visual contact with nature impact on health and well-being? *International Journal of Environmental Research and Public Health, 6*(9), 2332–43.

Groenewegen, P. P., van den Berg, A. E., de Vries, S., & Verheij, R. A. (2006). Vitamin G: Effects of green space on health, well-being, and social safety. *BMC Public Health, 6,* 149.

Groenewegen, P. P., van den Berg, A. E., Maas, J., Verheij, R. A., & de Vries, S. (2012). Is a green residential environment better for health? If so, why? *Annals of the Association of American Geographers, 102*(5), 996–1003.

Hansen-Ketchum, P. A., & Halpenny, E. A. (2011). Engaging with nature to promote health: Bridging research silos to examine the evidence. *Health Promotion International, 26*(1), 100–8.

Harnik, P., & Kimball, A. (2005). If you don't count, your park won't count. *Parks and Recreation, 40*(6), 8–14.

Hartig, T. (1993). Nature experience in transactional perspective. *Landscape and Urban Planning, 25*(1–2), 17–36.

Hartig, T., Mitchell, R., de Vries, S., & Frumkin, H. (2014). Nature and health. *Annual Review of Public Health, 35*, 21.1–21.22.

Honari, M., & Boleyn, T. (1999). *Health ecology: Health, culture, and human–environment interaction*. London: Routledge.

Jackson, L. E., Daniel, J., McCorkle, B., Sears, A., & Bush, K. F. (2013). Linking ecosystem services and human health: The Eco-Health Relationship Browser. *International Journal of Public Health, 58*(5), 747–55.

Kellert, S. R. (2005). *Building for life: Designing and understanding the human–nature connection*. Washington, DC: Island Press.

Kellert, S. R., & Wilson, E. O. (1993). *The biophilia hypothesis*. Washington, DC: Island Press.

Kellert, S. R., Heerwagen, J., & Mador, M. (2008). *Biophilic design: The theory, science and practice of bringing buildings to life*. New York, NY: John Wiley.

Khoo Teck Puat Hospital. (2015). A healing environment. Retrieved January 30, 2015, from https://www.ktph.com.sg/main/explore_ktph_pages/232/A_Healing_Environment

Land Trust Alliance. (2012). Voters approve 81% of land conservation ballot measures. Retrieved April 2, 2014, from www.landtrustalliance.org/policy/public-funding/voters-enthusiastically-approve-new-spending-on-conservation-nationwide

Laughlin, D., & Black, S. (Eds.). (1995). *Poverty and health: Tools for change*. Birmingham: Baring Foundation.

Li, Q., Morimoto, K., Kobayashi, M., Inagaki, H., Katsumata, M., Hirata, Y., et al. (2008a). Visiting a forest, but not a city, increases human natural killer activity and expression of anti-cancer proteins. *International Journal of Immunopathology and Pharmacology, 21*(1), 117–27.

Li, Q., Morimoto, K., Kobayashi, M., Inagaki, H., Katsumata, M., Hirata, Y., et al. (2008b). A forest bathing trip increases human natural killer activity and expression of anti-cancer proteins in female subjects. *Journal of Biological Regulators and Homeostatic Agents*, *22*(1), 45–55.

Lindheim, R., & Syme, S. (1983). Environments, people, and health. *Annual Review of Public Health*, *4*, 335–59.

Lindsey, G., & Knaap, G. (1999). Willingness to pay for urban greenway projects. *Journal of the American Planning Association*, *65*(3), 297–313.

Little, C. E. (1989). *Greenways for America*. Baltimore, MD: Johns Hopkins University Press.

Liu, J., Dietz, T., Carpenter, S., & Folke, C. (2007). Coupled human and natural systems. *Ambio*, *36*(8), 639–49.

Liu, J., Dietz, T., Carpenter, S. R., Alberti, M., Folke, C., Moran, E., et al. (2007). Complexity of coupled human and natural systems. *Science*, *317*(5844), 1513–16.

Louv, R. (2011). *The nature principle: Human restoration and the end of nature-deficit disorder*. Chapel Hill, NC: Algonquin Books.

Lovell, S. T., & Johnston, D. M. (2009). Creating multifunctional landscapes: How can the field of ecology inform the design of the landscape? *Frontiers in Ecology and the Environment*, *7*(4), 212–20.

Maas, J., Verheij, R. A., Groenewegen, P. P., de Vries, S., & Spreeuwenberg, P. (2006). Green space, urbanity, and health: How strong is the relation? *Journal of Epidemiology and Community Health*, *60*(7), 587–92.

Maas, J., Verheij, R. A., de Vries, S., Spreeuwenberg, P., Schellevis, F. G., & Groenewegen, P. P. (2009). Morbidity is related to a green living environment. *Journal of Epidemiology and Community Health*, *63*(12), 967–73.

Maslow, A. H. (1943). A theory of human motivation. *Psychological Review*, *50*, 370–96.

McConnell, W. J., Millington, J. D. A., Reo, N. J., Alberti, M., Asbjornsen, H., Baker, L. A., et al. (2011). Research on coupled human and natural systems (CHANS): Approach, challenges, and strategies. *Bulletin of the Ecological Society of America*, *92*(2), 218–28.

Melillo, J., & Sala, O. (2008). Ecosystem services. In E. Chivian & A. Bernstein (Eds.), *Sustaining life: How human health depends on biodiversity* (pp. 75–115). New York, NY: Oxford University Press.

Millennium Ecosystem Assessment. (2005a). *Ecosystems and human well-being: Health synthesis*. Geneva: WHO.

Millennium Ecosystem Assessment. (2005b). *Ecosystems and human well-being: Synthesis.* Washington, DC: Island Press.

Mitchell, R., & Popham, F. (2008). Effect of exposure to natural environment on health inequalities: An observational population study. *The Lancet, 372*(9650), 1655–60.

Myers, S. S., Gaffikin, L., Golden, C. D., Ostfeld, R. S., Redford, K. H., Ricketts, T. H., et al. (2013). Human health impacts of ecosystem alteration. *Proceedings of the National Academy of Sciences, 110*(47), 18753–60.

Niemelä, J., Saarela, S.-R., Söderman, T., Kopperoinen, L., Yli-Pelkonen, V., Väre, S., & Kotze, D. J. (2010). Using the ecosystem services approach for better planning and conservation of urban green spaces: A Finland case study. *Biodiversity and Conservation, 19*(11), 3225–43.

Nisbet, E. K., Zelenski, J. M., & Murphy, S. A. (2011). Happiness is in our nature: Exploring nature relatedness as a contributor to subjective well-being. *Journal of Happiness Studies, 12*(2), 303–22.

Ohtsuka, Y., Yabunaka, N., & Takayama, S. (1998). Shinrin-yoku (forest-air bathing and walking) effectively decreases blood glucose levels in diabetic patients. *International Journal of Biometeorology, 41*(3), 125–7.

Parkes, M., Panelli, R., & Weinstein, P. (2003). Converging paradigms for environmental health theory and practice. *Environmental Health Perspectives, 111*(5), 669–75.

Pretty, J., Griffin, M., Peacock, J., Hine, R., Sellens, M., & South, N. (2005). *A countryside for health and wellbeing: The physical and mental health benefits of green exercise.* A report for the Countryside Recreation Network, UK.

Prüss-Ustün, A., Bonjour, S., & Corvalán, C. (2008). The impact of the environment on health by country: A meta-synthesis. *Environmental Health, 7,* 7.

Richardson, E. A., Mitchell, R., Hartig, T., de Vries, S., Astell-Burt, T., & Frumkin, H. (2012). Green cities and health: A question of scale? *Journal of Epidemiology and Community Health, 66*(2), 160–5.

Rouse, D. C., & Bunster-Ossa, I. F. (2013). *Green infrastructure: A landscape approach.* Chicago, IL: APA Planners Press.

Sælensminde, K. (2004). Cost–benefit analyses of walking and cycling track networks taking into account insecurity, health effects and external costs of motorized traffic. *Transportation Research Part A: Policy and Practice, 38*(8), 593–606.

Sijtsma, F. J., de Vries, S., van Hinsberg, A., & Diederiks, J. (2012). Does "grey" urban living lead to more "green" holiday nights? A Netherlands case study. *Landscape and Urban Planning, 105*(3), 250–7.

Slater, P. (1970). *The pursuit of loneliness: American culture at the breaking point.* Boston, MA: Beacon Press.

Stigsdotter, U. K., Ekholm, O., Schipperijn, J., Toftager, M., Kamper-Jørgensen, F., & Randrup, T. B. (2010). Health promoting outdoor environments—associations between green space, and health, health-related quality of life and stress based on a Danish national representative survey. *Scandinavian Journal of Public Health, 38*(4), 411–17.

Stokes, D. E. (1997). *Pasteur's quadrant: Basic science and technological innovation.* Washington, DC: Brookings Institution Press.

Takano, T., Nakamura, K., & Watanabe, M. (2002). Urban residential environments and senior citizens' longevity in megacity areas: The importance of walkable green spaces. *Journal of Epidemiology and Community Health, 56*(12), 913–18.

The Economics of Ecosystems and Biodiversity. (n.d.). Ecosystem services. Retrieved May 22, 2015, from www.teebweb.org/resources/ecosystem-services/

Ulrich, R. S. (1993). Biophilia, biophobia, and natural landscapes. In S. Kellert & E. O. Wilson (Eds.), *The biophilia hypothesis* (pp. 73–137). Washington, DC: Island Press.

UNEP. (2006). *Marine and coastal ecosystems and human well-being: A synthesis report based on the findings of the Millennium Ecosystem Assessment.* Nairobi, Kenya: United Nations Environment Programme.

van Essen, H., Schroten, A., Otten, M., Sutter, D., Schreyer, C., Zandonella, R., et al. (2011). *External costs of transport in Europe.* Delft, the Netherlands: CE Delft, Infras, Fraunhofer ISI.

Van Herzele, A., Bell, S., Hartig, T., Podesta, M., & van Zon, R. (2011). Health benefits of nature experience: The challenge of linking practice and research. In K. Nilsson, M. Sangster, C. Gallis, T. Hartig, S. de Vries, K. Seeland, & J. Schipperijn (Eds.), *Forests, trees and human health* (pp. 169–82). New York, NY: Springer.

Velarde, M. D., Fry, G., & Tveit, M. (2007). Health effects of viewing landscapes: Landscape types in environmental psychology. *Urban Forestry & Urban Greening, 6*(4), 199–212.

Ward Thompson, C., Roe, J., Aspinall, P., Mitchell, R., Clow, A., & Miller, D. (2012). More green space is linked to less stress in deprived communities: Evidence from salivary cortisol patterns. *Landscape and Urban Planning, 105*(3), 221–9.

Wegner, G., & Pascual, U. (2011). Cost–benefit analysis in the context of ecosystem services for human well-being: A multidisciplinary critique. *Global Environmental Change, 21*(2), 492–504.

White, M. P., Alcock, I., Wheeler, B. W., & Depledge, M. H. (2013). Would you be happier living in a greener urban area? A fixed-effects analysis of panel data. *Psychological Science*, (April), 1–9.

Wilcox, B. A., Aguirre, A., Daszak, P., Horwitz, P., Martens, P., Parkes, M., et al. (2004). EcoHealth: A transdisciplinary imperative for a sustainable future. *EcoHealth, 1*(1), 3–5.

Williams, A. (1999). *Therapeutic landscapes: The dynamic between place and wellness.* Lanham, MD: University Press of America.

Wilson, E. O. (1984). *Biophilia*. Cambridge, MA: Harvard University Press.

Wolf, K. L. (2014). Water and wellness: Green infrastructure for health co-benefits. *Stormwater Report*, (April), 1–4.

Wright, H. (2011). Understanding green infrastructure: The development of a contested concept in England. *Local Environment, 16*(10), 1003–19.

Chapter Two

The evolution of the ecology of health

The significance given to the role of the physical environment in supporting health has waxed and waned over eons of recorded history. Likewise, so too has the significance of the natural environment and efforts to restore and conserve it for the purpose of supporting health. This ebb and flow is represented in the evolving paradigms of the ecological influences on health. Public health has progressed from considerations of the physical environment as a static external entity to a contemporary ecological paradigm with dynamic human–environment interactions. The now decades-old "new" public health and full embrace of ecological influences has created the theoretical space for the natural environment to be considered as fundamental to health once again. Current ecological models of health prominently feature the natural environment and biodiversity and ecosystems. Recognizing the encompassing and permeating role of the natural environment on health not only gives credence to the ample historical examples of efforts to protect the natural environment for the purpose of improving health but also sets the stage for a more prominent role for GI as the human life-support system.[1]

The following three sections proceed first with a very brief tour of the evolution of ecological thought in public health. This provides some context to the second section that recounts how the natural environment was considered essential for health in the period when manipulating the environment—later distinguished as urban, city, town, or spatial planning—and public health were indistinguishable. The third section shows the steady increase in the prominence of the natural environment in ecological models of health in the era of the "new" public health.

The evolution of an ecological conception of health

Recognizing that human health is affected by a complex web of determinants and processes, including the reciprocal determinism between the environment and human actions, represents an ecological way of thinking about health. This ecological conception of health is not new (Duhl & Sanchez, 1999).[2] It is "at least since the time of Hippocrates's essay 'Air, Water, and Places,' [that] humans have been aware of the many connections between health and the Environment" (McCally, 2002, p. 1). The ancient Greeks considered the environment critical to health but were largely concerned with finding the "right" environment and not necessarily manipulating the environment to make it "right" for health. Following the ancient Greeks in an era of nonspecific sanitation, the Romans, with their notable emphasis on civil engineering, were not searching for the "right" environment but instead attempting mastery over it to create an environment conducive to health. Extending through the medieval period and lasting until germ theory came to prominence in the later nineteenth century, it was the control of miasma—or foul, pathogenic air—that was considered essential to protecting health. The presence of miasma was one of many factors thought to affect the salubrity or wellness of the local environment. "Any major environmental element—land form, water moving and still, climate patterns, vegetation, wind patterns, history of local epidemics—had its role to play in whether or not an observer assessed a site as salubrious or not" (Martensen, 2009, p. 28). It is with salubrity that we witness a harkening back to the Greeks and the quest for the "right" kind of environment to support health, with salubrity being heavily influenced by the natural environment and processes.

The miasmists and germ theorists, although for different reasons, were working under an ecological model of public health: the miasmists because they believed that persons of low social station and character were ill because of their increased, and what they sometimes considered deserved, exposure to foul air, and the germ theorists because they recognized that social actions, such as poor sanitation, led to exposure to disease-causing pathogens. The discovery of the existence of microscopic organisms that could make us sick living all around and on us was revolutionary, but the revolution was not immediately decisive.

The clash between the miasmists and germ theorists was evident in one of the most famous cases of early epidemiology. Dr. John Snow's investigation into the London cholera epidemic of 1854 (Johnson, 2007) was a blow to the miasmists that eventually led to the victory of germ theory. While this proved to be an advancement in the protection of humans from infectious (communicable) disease, it did not fare well for an ecological health paradigm. As is often the case if all you have is a hammer, everything looks like a nail (Maslow, 1966), the discovery of pathogens eventually led public health into an era of immunizing people against these pathogens with a reduced focus on the environments where pathogens thrived and the ways in which humans came into contact with them. Considering the resources devoted to clinical health care and pharmaceutical solutions today, the dominant paradigm is still one of identifying the biological causes of disease with far less resources devoted to the "upstream" environmental, social, and behavioral factors that are eventually evidenced in biologically observable conditions.

We see early hints of the questioning of the biomedical paradigm and a return to ecological thinking when Duhl (1963) states "our old [biomedical] methods of seeking solutions were inadequate and have to be replaced by far more complex and sophisticated perceptions of the interplay between man [i.e. people] and his [i.e. their] environment" (p. 1). The McKeown hypothesis brought data to bear that the emphasis on medical treatment was not having the desired effect of reducing the disease burden in populations (McKeown, 1971, 1979). In fact, with the exception of the small pox vaccination (with its important but overall very small effect on reducing mortality), McKeown demonstrated that it was behavioral and environmental factors that accounted for reduced mortality in the age of immunization, clinical care, and medicine in England and Wales between 1840 and 1970. In line with Darwinian evolution now firmly ingrained in the scientific community, McKeown also noted the evolutionary connection between humans and their environment, noting that it was quite possibly a divergence from evolutionary adapted environments that exacerbated disease. With the natural environment dominating almost the entirety of human evolution, it was only a small step then to consider the role of the natural environment in human health and the degradation and neglect of this type of environment that could lead to poor health.

In an attempt to remedy the shortcomings of the biomedical paradigm, we see ecological thinking reflected in the "new" public health. The reacquaintance with the environment, in what had yet to be termed the "new" public health, was evident in the Lalonde Report, *A New Perspective on the Health of Canadians* (Lalonde, 1974). At the time of the Lalonde Report, raising the environment to the same level as health care was a "radical step" (Morris, Beck, Hanlon, &

Robertson, 2006, p. 891). It was just over a decade after the Lalonde Report, in 1986, when the charter of the new public health was delivered at the First International Conference on Health Promotion in Ottawa, Canada. In the Ottawa Charter it was declared that "the fundamental conditions and resources for health are peace, shelter, education, food, income, a stable ecosystem, sustainable resources, social justice and equity. Improvement in health requires a secure foundation in these basic prerequisites" (WHO, n.d.).[3] This laudably comprehensive list explicitly credits the reliance on stable ecosystems and sustainable resources as "fundamental conditions" for health.[4, 5] Implicit is the fundamental importance of the natural environment in supporting these conditions, but the natural environment is also critical to the other listed fundamental conditions and resources. Without the GI that supports a healthy ecosystem, many of the other prerequisites of health are simply undeliverable. Peace is often threatened by conflicts over the control of natural resources,[6] shelter is built with nature's raw and manufactured materials, food relies on the soil, and income (closely related to the other stated conditions of education and equity) is derived from the direct reaping of natural resources or the conversion of these resources into marketable products. This is not to suggest that the role of the natural environment in the delivery of these conditions was lost on the "new" public health. On the contrary, it was a time when recounting the ecology of these connections was considered essential to advancing the field.

> Practitioners of the new [post-1986] public health need to have a good grounding in ecology and a vision of how to reconcile the natural and the built environments. They need to revisit all the topics of the old public health—housing, food, water, sanitation, education, occupation, transportation, genetics and microbiology, and medical and social services—and reexamine them with ecological eyes.
>
> (Ashton, 1991, p. 190)

This reconciliation of the natural and built environments noted by Ashton (1991) is exactly what GI provides, and it is this reconciliation that is essential to the delivery of the ecosystem services, but a truly ecological model of health is not complete without two other considerations, one being the interactions between organisms and the other being the influence those organisms have on their environment. The first takes center stage in aptly named socio-ecological models, and the second is central to the practice of spatial planning (i.e. urban and landscape planning).

The application of the ecological paradigm in public health has largely been a socio-ecological approach with the central organism being human and the focus being on the relationships within and between human communities. Based on the pioneering work by Urie Bronfenbrenner in his theory of human development, socio-ecological models acknowledge the environment, but the environment in question is the social environment composed of human actors, social systems, and institutions (Figure 2.1) (Bronfenbrenner, 1977, 1979, 1994). Many decades of research have led to many refinements of Bronfrenbrenner's model over the years; this socio-ecological approach has proved extremely useful in understanding the social determinants of health. It is socio-ecological in that there are interactions

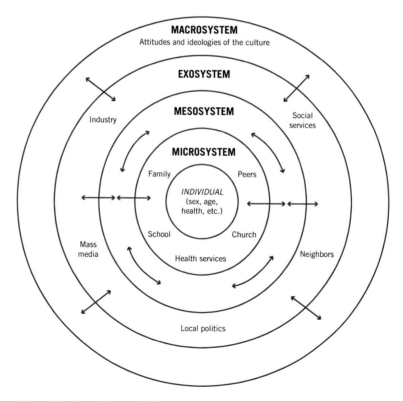

Figure 2.1: Bronfenbrenner's Ecological Theory of Development
Source: Hchokr.

between humans occurring at various scales of social organization. The ecology stems from the social actors and institutions being external to the individual. What is missing is a consideration of the physical environment that results from the interactions and activities in the social environment. What this socio-ecological paradigm also denotes is one of the greatest distinguishing factors between public health and medicine. This is the preventive nature of public health focused on the "upstream" or root causes of disease and also the application of preventive measures at the group or community level. While it is the individual that benefits from public health intervention, it is the designing of interventions for groups of individuals with common risks that puts the "public" in public health.

The succession of models presented in the upcoming "Greening of the Ecological Public Health Paradigm" section build on the socio-ecological approach but add the physical environment construct. The natural environment, as one form of the physical environment, is the context in which all other social and physical spheres are nested. The greener models are considered more fully "ecological" and not strictly "socio-ecological." Granted, the approach is still anthropocentric in that these models are focused on human health, humans are the central organism in question, and it is the human interaction with the natural environment (and GI) that we aim to understand the health benefits of. While the addition of the physical, natural environment brings us closer to conforming to the terse but densely packed definition of ecology, human health is still the central concern so these models are aptly named "ecological models of health" and not simply ecological models.

Even with the addition of the physical environment, there are still shortcomings in ecological models of health maintaining a true ecological perspective. A socio-ecological approach accounts for the interactions between human organisms, and the addition of the physical environment accounts for its influence on human health, but an ecological model of health also needs to capture the reciprocity between humans and the environment. This involves not only the influence of the social and physical environments on human health but also the influence of humans on the environmental elements necessary to maintain health. Human actions that influence the natural environment need to be considered in a complete ecological model of health.[7] The environmental sciences have made enormous advances in uncovering the natural environment elements and relationships necessary for viable ecosystems with the largest threat to these ecosystems being human actions. Protecting the places from which health benefits are derived is evident in an emerging socio-ecological model of sustainability (Maller, Townsend, Pryor, Brown, & St Leger, 2006) where human actions that threaten environmental sustainability are also viewed as threatening human health and well-being. This is expressed much more eloquently by Wendell Berry (1993, p. 14) when he states

> If we speak of a healthy community, we cannot be speaking of a community that is merely human. We are talking about a neighborhood of humans in a place, plus the place itself: its soil, its water, its air, and all the families and tribes of the nonhuman creatures that belong to it. What is more, it is only if this whole community is healthy ... [and] the human economy is in practical harmony with the nature of the place, that its members can remain healthy and be healthy in body and mind and live in a sustainable manner.

Recognizing the influence humans have on the environment and its ability to continue to support health leads us to spatial planning and the practice of environmental manipulation. Next, we briefly visit the history of how altering environments was believed to be important to health and examine the many efforts to include nature in built environments for the benefit of health and well-being. Many successful efforts to conserve the natural environment are evidenced in the cities and landscapes that humans currently occupy. There were many pioneers whose actions to preserve GI are still benefiting the health of those that inhabit the environments touched by their vision.

A brief history of spatial planning for health

Ecological thinking and shifting attitudes towards the environment in public health have occurred concurrently with urban development. Throughout this development there has been a recognition of the health benefits of bringing GI (although not called GI until quite recently) into cities. This connection extends from settlements in the Fertile Crescent with enclosed parks or orchards, *pairidaeza* or "paradise" (Ward Thompson, 2011, p. 188) and Chaldean civilizations (Williams, 1944) to medieval landscapes in Chinese merchant gardens (Hongxun, 1982). It was approaching the late medieval period when cities grew to such a size that urban dwellers could ostensibly be removed from any exposure to the natural environment, but it was not until the industrial revolution that we witness a concerted effort in Europe to include representations of nature in the urban environment for the explicit purpose of exposing urban inhabitants to fleeting and degraded nature and its restorative and prophylactic properties. The language and proposals of the time extol the need to incorporate GI into human settlements to support health.

This period has often been lauded as a turning point when much needed urban gray infrastructure was introduced to protect public health, and urban planning and public health were essentially inseparable (Hebbert, 1999). The public health and urban reforms of this era are well documented in the literature (Ashton & Ubido, 1991), particularly the literature over the past two decades when there has been a renewed interest in the intersection between urban planning and public health, but it was the eventual and unignorable by-products of the industrial revolution that brought about the pressure to clean up the environment.

There were both "brown and green" motivations for improving the conditions in industrial-era cities (Olival, Hoguet, & Daszak, 2013). The brown motivation involved the control of pollution, with "brown" in this case being human waste and the by-products of the industrial revolution. This is also a time when there are also "green" motivations for protecting and rehabilitating the environment for the purposes of protecting public health. The green motivation was inspired by a recognition that "nature may heal and give strength to body and soul alike" (Muir, 1912, p. 256). The notion that the protection of the natural landscape was essential to human health and development was firmly taking hold as both an urban planning and a land-use principle. The perceived wellness-enhancing qualities of the natural landscape were a driving force behind nineteenth-century landscape architecture and the urban parks movement aimed at bringing greenspace into cities. Parks were perceived as a defense against what were believed to be, at the time, the miasma-induced plagues of rapidly growing and horribly unhygienic cities and as a respite that could induce heightened character and moral refinement (Walker & Duffield, 1983; Ward Thompson, 1998; Warpole, 2007). It is also during this time when the green is considered for its ability to remediate the brown. Urban parks were touted as the green "lungs" of the polluted city. This idea began in the industrial UK and bolstered arguments for parks in all of the world's major cities of the time (Ward Thompson, 2011).

Frederick Law Olmsted is credited with having the greatest singular influence on the urban parks movement in North America, and health was an explicit motivation for his desire to introduce parks into rapidly expanding American cities (Figure 2.2). His view on the health benefits of parks mirrored that of the

European parks movement where greenspace was recognized as a mental respite but also a resource "to counter disease and physical ill-health" (Ward Thompson, 2011, p. 192). Lesser known, but also visionary, was John Rauch. His report, *Public Parks: Their Effects upon the Moral, Physical and Sanitary Condition of the Inhabitants of Large Cities; with Special Reference to the City of Chicago* (1869), made the explicit connection between city greening and health. It instigated massive park projects and the planting of thousands of trees throughout Chicago. While Olmsted and Rauch were moving to increase contact with nature in cities, there was also a movement to reap the health benefits of nature by fleeing the city.

In his exploration into the rise of suburban *Bourgeois Utopias*, Fishman (1989) notes the prevailing attitude of the time, "only if urban dwellers can experience a daily 'change both of scene and air' will their physical, psychological, and moral health be maintained, and the full benefits of the city for civilization preserved" (p. 127). Well before the post-World War II popularity of the suburbs—that ironically would lead to urban sprawl and the large-scale and harmful fragmentation of the

Figure 2.2: Central Park (just left of center) in New York City, a Frederick Law Olmsted creation

Source: US National Aeronautical and Space Administration.

landscape—near the turn of the twentieth century in the UK there were grand plans being made for the creation of cities that would strike the right balance of development and conservation believed to be necessary to maintain health and harmony.

An example of one of these grand plans is Richardson's (1876) proposal for a utopian community (Hygeia) that was sensitive to natural processes and incorporated natural elements for the purpose of supporting health.[8] Following Richardson, the best known example came in 1898 when Ebenezer Howard proposed his grand plans for Garden Cities that reinstalled the critical link between urban development and personal and communal health and well-being (Howard, 1965, 2009).[9] Howard had witnessed how the industrial revolution in the UK, and later across Europe, had caused major changes in how people lived. Population densities increased significantly with rapid migration to urban areas. A lack of adequate gray infrastructure and GI to support urban populations led to major public health problems. Communicative disease rates increased as the prevalence of cholera, diarrhea, and bronchial viruses spread across cities and worker settlements.

Howard's "three-magnets" (Figure 2.3) approach to integrating the civic duties of the town with the accessibility and amenity value of the country restructured the ways that urban areas could be built. Letchworth and Welwyn were the first city experiments specifically planned to integrate Howard's Garden City principles (Figure 2.4). Both cities were designed to strategically invest in accessible public space close to homes and centers of employment. This, as Howard observed, provided the physical framework of spaces that would, thorough proximity and accessibility, encourage people to spend more time in public greenspaces. Howard's proposals offered a preventive approach to public health that balanced a sufficiently proportional public realm alongside adequate housing to minimize the risk of spreading communicative disease (Hall, 2002).

Current development policy in the UK is now revisiting Howard's Garden Cities ideals. Strong advocacy from the Town and Country Planning Association has informed government that the principles proposed by Howard and his successors (Fishman, 1982) are as relevant today as they were in 1898 (Town and Country Planning Association, 2012a). Providing spaces that promote outdoor living, social interaction, and an interest in the natural environment are principles that promote a positive relationship between the natural environment and health (Town and Country Planning Association, 2012b). Today, we would consider

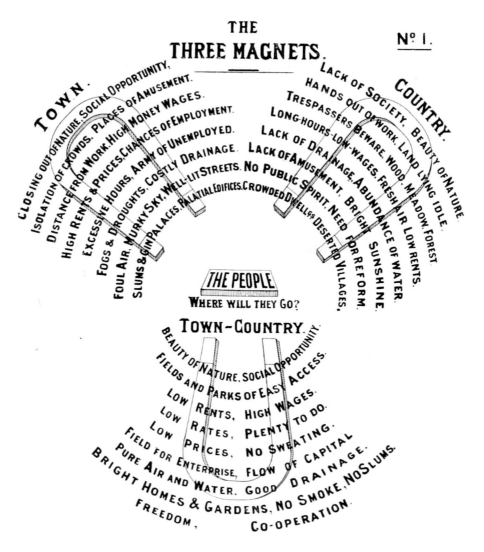

Figure 2.3: Ebenezer Howard's Three Magnets
Source: Ebenezer Howard.

Howard's countryside weaved within the urban fabric the GI of a city. This GI approach is well beyond the inclusion of parks and gardens in Victorian-era urban reforms aimed at providing respite. A GI approach recognizes larger considerations of the landscape components necessary for functioning ecosystems as the human

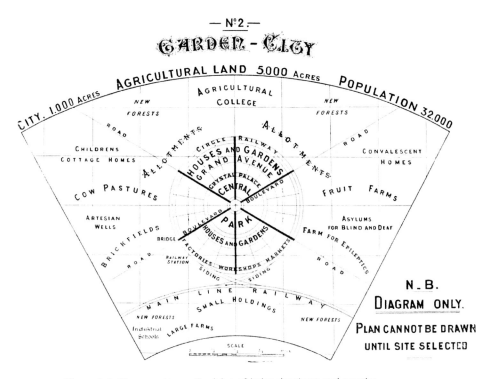

Figure 2.4: Ebenezer Howard's vision of balancing town and country
Source: Ebenezer Howard.

life-support system. Our understanding of the role for urban parks in supporting biophysical processes and human needs has come a long way, and its importance in serving this function should remain a priority (Ward Thompson, 1998). In fact, targets have been set by the World Health Organization that each city dweller needs nine square meters[10] of greenspace to reap recreational benefits as well as other health benefits stemming from GI (Brown & Grant, 2005; Niemelä et al., 2010).

It was in Howard's time when the distinction was drawn between urban planning and public health. Urban issues related to the increasingly squalid and unsanitary conditions in rapidly expanding cities instigated many public health reforms that included plans to improve the physical design of cities to cope with environmentally determined disease. Architects, public health officials, and social workers all played a role in city conditions. Architects focused on place, and public health and social workers focused on people. It was the recognized need to address the overlap of place and people that formed the core of what was to become urban planning. This is not to say that planning has not struggled with the false dichotomy of place and people over the years. In fact, when modern urban planning was formalized at the very first urban planning conference in 1898,[11] an underpinning question was whether planning was about the physical design of places or about making places better for people (Erickson, 2012). Just as it was then, GI is about doing both.

On the surface it may appear that a GI approach is simply revisiting Howard's vision of a century ago, but GI reflects a much more sophisticated understanding of just how critical the natural environment is to human health and well-being. This increased sophistication is reflected in the evolution of increasingly green ecological models of health.

The greening of the ecological public health paradigm

Despite the presence, and then prominence, of the natural environment in ecological models of health over the past 30 years, the natural environment has received far less attention than other constructs. This is true both in research investigating the ecological determinants of disease and also in the practice of altering environments to make them more supportive of health. This is not to say that the environment has been ignored. There have been tremendous advances made in understanding the risks associated with exposure to toxic elements in the environment, but we still know relatively little about how the planning and design of the physical environment and GI facilitates or hinders health and health behaviors. This has changed over the past decade with a renewed appreciation in research and practice for the complex ecological nature of the interdependence of human health and the natural environment and ecosystems.

In the progression of models that follow, we see the natural environment featured among other determinants and spheres of influence in an ecological way with implicit and explicit interrelationships between the social environment and the physical environment. A number of ecological models present all other ecological spheres (e.g. the built environment, community, and economy) as nested within and dependent upon the natural environment and biosphere. Later models provide us with more nuance on how alterations to the natural environment affects health. These frameworks are not, nor were they meant to be, comprehensive assemblages of the evidence on the determinants of health. Rather, they offer a way to organize the evidence to reveal implications and interdependencies. It is these types of models which can aid in the understanding and management of complex ecological systems (Heemskerk, 2003) and their influence on health.[12, 13]

Presented in Hancock and Perkins (1985) and modified slightly in Hancock (1985, 1986), the Mandala of Health (Figure 2.5) is among the earliest recognitions, in the era of the new public health, of the natural environment and biosphere in a bio–psycho–social–environmental conceptual model.[14] It communicates the human-made (built) environment as encompassed by the natural environment and processes occurring in our biosphere. With the biosphere encompassing all other constructs in the model, its prominence is clear, and with the nesting of all other spheres and constructs within the biosphere we can assume that there

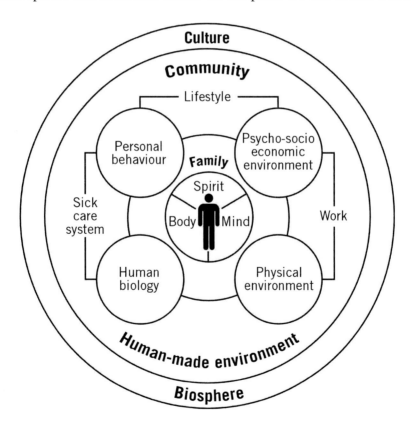

Figure 2.5: The Mandala of Health
Source: Hancock (1986).

are interactions between the biosphere and other determinants. Culture is also presented here, but this notion of culture extends beyond cultural differences that may exist at the community level, such as in a neighborhood. An example of culture at this level could be a "Western, democratic, technologic, science-based, Judeo-Christian culture" (Hancock, 1985, p. 3). Viewing the world through this macro-level cultural lens could certainly influence not only how people perceive health but also the attitudes towards the biosphere's role in supporting it. The overlapping forces of culture and biosphere are reminiscent of the philosophy of Lewis Mumford, one of the pillars in twentieth-century US urban planning. Mumford "consistently argued that the physical design of cities and their economic functions were secondary to their relationship to the natural environment and to the spiritual values of human community" (LeGates & Stout, 2011, p. 91). The physical environment in this model includes the built environment factors of housing and workplace conditions, but these are subservient to "the ultimate determinant of health," the biosphere (Hancock, 1985, p. 3).

Appearing some years later, the model by Evans and Stoddart (1990) shows the direction of influence that the physical environment can have on disease and how the physical environment might operate via individual response (Figure 2.6). It is not clear in this model what is included in the physical environment construct, but, inserting the natural environment here as a type of physical environment, the natural environment has a direct influence on disease and also indirectly on "health and function" via individual response. Health and function influence prosperity that then circles back on the physical environment. With this feedback, the model is ecological, but an appreciation for natural environment can only be assumed. This varies from the Hancock (1986) model and others to come where the natural environment is an overarching force. Although not a "green" model, this model was an advancement because it elucidated how health care was not the solution to public health, and it created the conceptual space for the significant role of physical environment in the prevention of disease.

Nearly a decade later, the Butterfly Model of Health (VanLeeuwen, Waltner-Toews, Abernathy, & Smit, 1999) provides another useful conceptualization of the influence of the physical, this time the biophysical, environment on health (Figure 2.7). In the tradition of the many evolving models of health presented here and reviewed by VanLeeuwen et al. (1999), the Butterfly Model recognizes the crucial role of the socioeconomic environment, but it advances the biophysical (natural) environment as equally significant. Humans, and individual human characteristics

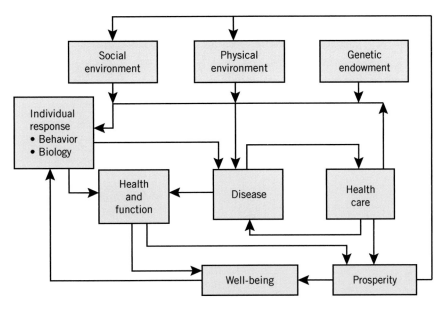

Figure 2.6: Ecological relationship of physical environment and health
Source: Evans and Stoddart (1990).

and behaviors, are nested at the intersection of the biophysical and socioeconomic environment. Humans are placed in an ecosystem context with reciprocity
between humans and the biophysical environment, recognizing that humans move
in and out of a variety of environments which they are both influenced by and
have an influence upon. This model more accurately represents the ecological
relationship between humans and the environment because it includes not only the
influence of the environment on human health but also the influence of humans
on the environmental elements necessary to maintain health. In this ecological
approach, health becomes a process rather than quantifiable outcome (Kickbusch,
1989) because measures of health status are constantly changing due to dynamic
human–nature interactions. This model also makes a distinction between the more
proximal environment and the "external" environment. Granted, all environments
are external, but the "external" distinction here reflects the degree of externality.
These varying degrees of externality have been parsed out in other models as
intermediate-, meso- or community-level, and as the fundamental macro-level
influence of the biophysical environment (Northridge, Sclar, & Biswas, 2003;
Schulz & Northridge, 2004).

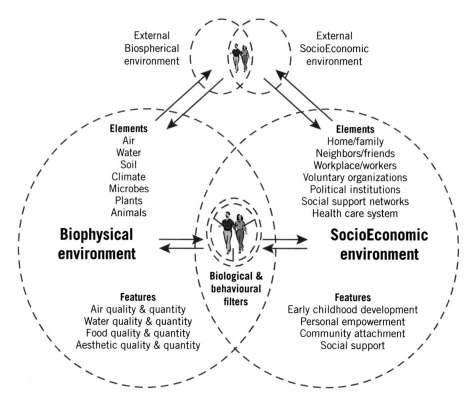

Figure 2.7: The Butterfly Model
Source: VanLeeuwen et al. (1999).

At about the same time as the Butterfly Model, we see the insertion of human social systems into ecological models coming from the direction of forestry and environmental studies. The *human ecosystem* approach to ecosystem management is an environmentally driven organizing concept that inserts social systems into the study of ecosystems and seeks to understand the reciprocal relationship between environmental quality and human quality of life (Force & Machlis, 1997; Machlis, Force, & Burch Jr, 1997). While the thrust of the human ecosystem approach is protecting the environment, the reciprocity between doing so and protecting human health makes human ecology highly complementary to ecological models of health. The only major difference between environmentally focused human ecology and contemporary models of health is the outcome of interest resulting from this interaction, one being sustainable natural systems and the other being human health. As I hope we can appreciate by this point, environmental protection is health promotion, and although human health is not always an explicit goal of environmental initiatives, human health is coupled with and benefits from them (Rapport, Gaudet, Constanza, Epstein, & Levins, 1998).

The Health Map (Figure 2.8) depicts the importance of the global ecosystem and the enveloping influence of the natural environment on a healthy human habitat (Barton & Grant, 2006; Dahlgren & Whitehead, 1991).[15] Because the Health Map depicts "human habitation," it has the flexibility to be applied at various scales. The natural environment here could refer to the neighborhood but also to larger regional systems, both of which ultimately influence and are dependent

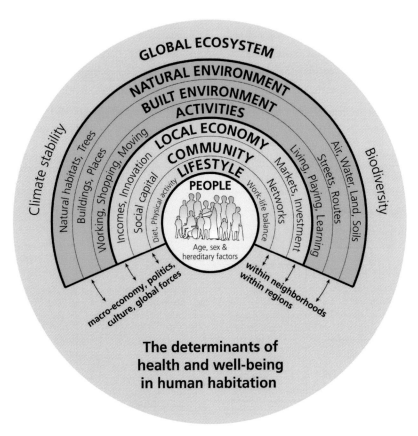

Figure 2.8: The Health Map

Source: Barton and Grant (2006) with attribution to Dahlgren and Whitehead (1991).

upon global ecosystems. Like the Butterfly Model, the human is an agent in the environment, community, and economy, and these spheres influence health through biological (hereditary) and behavioral lifestyle filters. The advancement in this model is an explicit recognition of climate stability and biodiversity as essential components of the macro-level global ecosystem. Although reciprocity is not explicit in the model, similar to the Mandala of Health, the nesting of all other spheres within the biosphere allows the interpreter of this model to assume that these smaller spheres also influence the biosphere. The resources and conditions in the smaller spheres mediate the relationship between health and natural environment by either insulating people from or making people more vulnerable to global environmental change. This insulating effect is explored more closely in Figure 2.11 (Myers & Patz, 2009).

A Public Health Ecology model (Figure 2.9) focuses on the interactions between the natural and built environments and specifically on the health outcomes that result from spatial planning decisions that alter the natural environment (Coutts, 2010).[16] Much like previous models, it explicitly advances the natural landscape as supporting health directly through environmental agents and stressors and indirectly through the behaviors that the environment facilitates

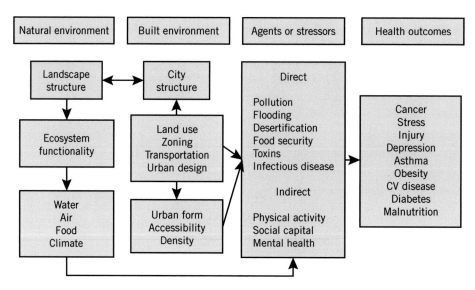

Figure 2.9: Public Health Ecology
Source: Coutts (2010).

or hinders. The presence or absence of elements of the natural environment can directly influence health. In one of the starkest examples of the direct human dependence on the natural environment, health (and life) would cease without a minimum quantity and quality of water and air. The structure of the natural environment and GI influences the ability of ecosystems to deliver ecosystem services in their necessary quality and abundance. Also, the natural environment can influence health indirectly through the behaviors that it facilitates or hinders.[17] Environmental barriers such as climate and the distance to and availability of natural resources are contributing factors to the choices people make—choices with health consequences. Unlike other models presented up to this point, Public Health Ecology advances the idea that the structure of the landscape and GI has an influence on the viability of ecosystems that can be threatened by urban development. In this and subsequent models, we begin to see how human alterations to the natural environment via construction of the built environment influence natural systems and health.

The Transformation via Balanced Exchanges (T-BE) model (Figure 2.10) presents the exchange of ecosystem services and human actions between the natural and human systems occurring in the built and natural environment contexts. It is similar to the Mandala of Health and Health Map in that the natural environment is the context in which the built environment is nested, but it reveals the interface of the human and natural systems within these contexts. Among the outcomes of the interactions between human and natural systems is human health. In TB-E, the human system is no longer at the center of the model but neither is the natural system. Instead, the human system is viewed in the context of the built environment and interacting and overlapping natural systems—the human and natural systems together acting as an integrated part of the larger global natural environment. The natural system encompasses ecoregions with ecosystems, communities, and organisms (Glaser et al., 2012). Ecological processes influence the ecological system but also the human system via ecosystem services. The built environment is shaped and planned by humans and contains structures such as buildings and infrastructure. At the interface of the human and natural system are the ecosystems along the urban–rural continuum (Niemelä, 1999; Pickett, Cadenasso, & Grove, 2001). These ecosystems form the urban–rural system which includes GI within the built environment. The human system includes individuals and relationships between individuals at the community and regional (or higher) levels. Humans not only depend on the environment via ecosystem services but also impact the environment with their actions. This in turn impacts human health.

A conceptualization of the health effects caused by human alterations to the landscape is found in Figure 2.11 (Myers & Patz, 2009). Global environmental changes in land-use or climate pose direct risks to health (e.g. ambient heat from urbanization) and also impair the ecosystem services on which health depends. A number of insulating layers of natural and social resources mediate the risks to health from these local and global drivers of environmental and ecosystem change. Examining these insulating layers from the outside in, and with a focus on land-use and land cover changes that diminish GI, it is evident that populations are at much greater risk of negative health outcomes if they have reached a threshold of resource constraint. It is likely that no significant changes to health would occur until a tipping point is reached. This is similar to the concept of natural capital, or our "savings account" of natural resources. We can continue for a while drawing

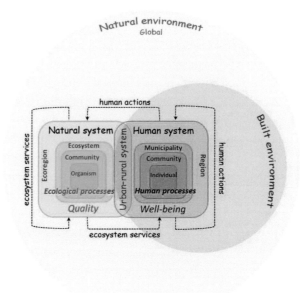

Figure 2.10: Transformation via Balanced Exchanges

Source: Coutts, Forkink, and Weiner (2014).

Note: Created by Annet Forkink. Adapted from Glaser et al. (2012); Millennium Ecosystem Assessment (2005); Pavao-Zuckerman (2000).

natural capital from the bank of the biosphere without replacing it, but this is sustainable only up to a certain threshold. There comes a point when nearly or fully exhausted natural capital will have health consequences. Until a threshold is reached, one could continue degrading natural capital in an unsustainable way, heading towards the threshold, but this threshold can be moved by extending the reach of locally diminished natural capital.

The next layer captures this extended reach by taking into account the fact that one can replace locally unavailable or depleted ecosystem services with those from other regions. The ability to supplement locally unavailable resources, due to the lack of natural occurrence or human depletion, can mediate the influence on health. The Ecological Transition, discussed shortly, reveals how the ability to do this is influenced by socioeconomic factors. When considering global resources, such as the atmosphere, this substitution is being done constantly. The largest polluters commit local offenses to health that, if contained, would quickly destroy the ecosystem services on which health depends, but much of the pollution that

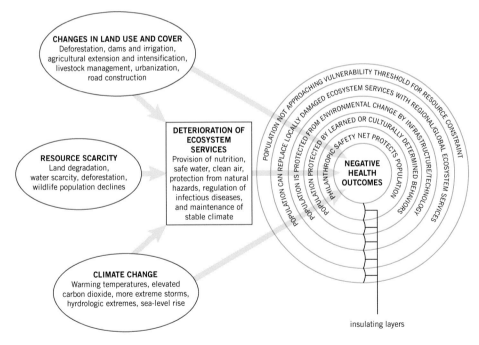

Figure 2.11: Health and landscape change

Source: Myers et al. (2013).

poses threats to ecosystems is being dispersed regionally and globally. Populations also supplant and augment services with physical infrastructure. Again, the Ecological Transition reveals how this may be an option for some, but it is an unfeasible solution for global health.

The next insulating layer reflects culturally determined behaviors that can reduce risks through adaptive behaviors. The importance of culture is explicit in the very first ecological model of health presented here, the Mandala of Health, it being on the same level as the biosphere. While there are culturally determined behaviors that influence the biosphere, culture is applied here to capture the learned behaviors that may become the social norms that help populations adapt to reductions in ecosystem services. These behaviors could involve treating water, planting more heat tolerant crops, and raising homes further from ground level to accommodate rising seas, storm surge, and inland floods. Thinking ecologically about the reciprocal influence of these behaviors on ecosystem services, these could also involve mitigation, such as the reduction of pollution through reforestation. Another mitigation action that could be added to this model is the effect of changes in land use (i.e. diminished GI) on climate change. The natural environment and ecosystems are changing without human interferences, but human disturbances to GI are having a dramatically accelerated effect on climatic systems and human health.

Finally, philanthropic safety nets can protect the health of those in areas where ecosystem services are diminished due to local mismanagement or regional or global forces outside of local control. What likely first comes to mind are sacks of grain being handed out off the back of a truck to persons suffering in drought-stricken countries, but philanthropy acts as a form of protection in more economically wealthy countries as well. A notable example of this was the international response to the effects of the earthquake and tsunami in Japan in 2011 that shifted Japan as one of the world's largest donors to one of its largest recipients of aid (Watts & Borger, 2011). The philanthropic safety net is not isolated to strictly the poor.

Adapting the environmental risk transition model (Smith & Ezzati, 2005) to include the potentially disproportionate health effects of ecological change, Myers et al. (2013) present their schematic of the Ecological Transition (Figure 2.12).

In this schematic, the population moves from a state (a) in which people rely primarily on natural systems for health-related ecosystem services to a state

(c) where they become reliant on engineered infrastructure and markets for these services while ecological systems get degraded over time. Over the course of this transition (b), there are numerous society-level mediating influences that are likely to change the differential vulnerabilities and health status of members of the population. These include equity of income distribution, type and strength of governance, philanthropic safety nets, characteristics of the natural environment etc. It is also likely that the health implications of such a transition will be different for different dimensions of health.

(Myers et al., 2013, p. 18756)

There is no question that socioeconomic conditions influence environmental sustainability. Two of the three pillars of sustainability (economics, equity) reflect their importance on the third (environment). What the Ecological Transition shows is that when the environment is threatened, economics will determine the degree of equity in whose health will suffer most. There is a universal human dependence on the natural environment regardless of economic characteristics, but

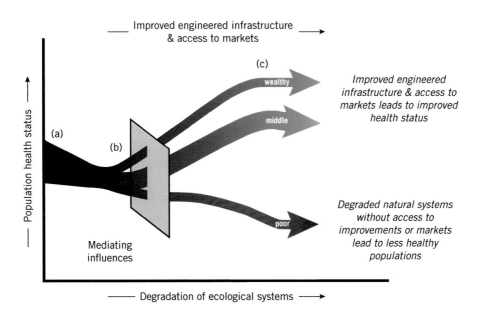

Figure 2.12: The Ecological Transition
Source: Myers et al. (2013).

these characteristics certainly alter the extent and type of health effects stemming from a reduction in ecosystem services. Those with the most resources are better able to cope with ecological change as they are better able to protect themselves with the gray infrastructure. This gray infrastructure compensates for the reduced capacity of ecosystems to provide essential services. This is not to say that the sole solution to a reduction in ecosystem services is to expand gray infrastructure. This adaptive strategy may work for a time, but the gray depends on the green. There are a number of ecosystem services that no current technology can replace, and there is no feasible way that enough gray infrastructure can be constructed to protect everyone at risk. There will be a point, beyond what is depicted in this model, at which the health status of high and middle income populations will decline as well, regardless of gray infrastructure or access to markets. Wealthier nations, as they shift their "heavy" pollutants to other countries and away from local view, are still responsible for, and will feel the impacts of, pollution with increased geographical scope (stresses on the biosphere) even though local environmental pollution and damage is less evident (McMichael et al., 2003, p. 4) (Figure 2.13). One strategy is to restore and protect the GI on which ecological systems rely to produce the ecosystem services on which all humans depend.

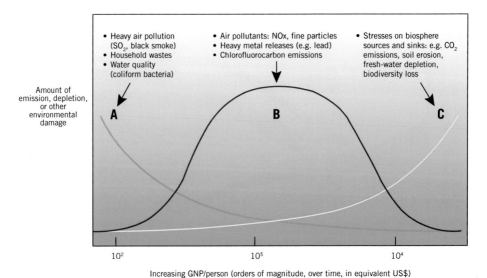

Figure 2.13: Environmental impacts on health resulting from human social and economic development

Source: McMichael et al. (2003).

The natural environment that supports ecological systems is absolutely necessary for humans to survive and thrive as a species, and by ignoring the reciprocal maintenance and determinism between humans and the natural environment (putting the burden on the shoulders of individual choice, economics), there is a risk of blaming the victim (the poor in the ecological transition) for health woes. The biomedical model focused on individual immunity and the provision of therapeutic care when immunity fails would have us blame poor health on individual failings, either in lifestyle choices or genetic factors. The individual is the centerpiece of an ecological model of health, but it is clear that the individual is limited in his/her ability to alter global, or even regional, forces that threaten local and individual health. Could we reasonably expect someone to implement proper sanitation in their daily lives in a sprawling informal settlement without the provision of sanitation infrastructure? Could we reasonably expect someone to eat healthier in an environment where healthy food options are nonexistent or economically out of reach? We blame the victim when we put the burden on individuals for their own diminished health in environments that cannot support health. It is the same for natural environment and GI. While it is a collection of individual choices that threaten the natural environment, so too are the health consequences of a diminished natural environment spread to all living things.

It is critical to note that fixing the environment or providing the "right" environment, as the Romans believed, is not a panacea for public health ills. Quite far from an environmentally deterministic view, the ecological paradigm makes explicit the role of the environment amidst, and influenced by, an array of other determinants of health. What is lacking in the current application of the ecological paradigm is the recognition of the full scope of the environment construct which recognizes our evolutionary and still fundamental need for the health-supporting benefits of the presence, access, and exposure to GI. We have moved beyond the Victorian-era belief that incorporating greenspace into urban environments will cure all urban mental and physical ills, but maybe we have moved too far. The evidence of the benefits of GI has continued to grow, but the application of this evidence in creating environments supportive of health has not.

Summary

The recognition of the physical environment as a determinant of health is nothing new; it is simply that its prominence in an ecological conception of health has shifted over eons of recorded history. The shifting in the prominence of the physical environment was played out in the practice of manipulating the physical environment for human habitation, or spatial planning. There was a time when planning and public health were one and the same, and the inclusion of both gray and green infrastructure in plans was public health practice. This shifted over the past half-century when our enchantment with biomedicine and individual immunity came at the expense of thinking ecologically about health. The reawakening to the role of the natural environment is evident in the evolution of ecological models of health in the era of the now decades-old "new" public health. In these models, the natural environment and biosphere form the foundation in which all other spheres in the ecological model are nested. Green infrastructure plays a crucial role in the ecological and biophysical processes of the biosphere that support life and health, but GI also permeates and supports physical and mental health derived from more socially oriented spheres and determinants. Despite the prominence of the natural environment in theoretical models, theory has yet to translate into an equally robust movement to conserve and protect GI for the benefit of protecting and promoting health.

Notes

1. A greatly condensed and early draft of this chapter was published by Coutts, Forkink, and Weiner (2014).
2. This assessment of public health thought is admittedly Western-centric and thus far from complete. This is in no way meant to discount the advances in civilization in other parts of the globe and the undoubted existence of considerations of the physical environment and evidence of human settlement infrastructure aimed at improving health. The story told here reflects solely what is accessible to the author.
3. It is in this same year that the World Health Organization's Healthy Cities initiative was born, a movement that harkens back to Victorian-era sanitary reforms (Ashton & Ubido, 1991).
4. The term "stable" to denote a viable ecosystem is no longer in vogue. It is now understood that a viable ecosystem may constantly be in a state of flux and that resilience is what is critical to sustained ecosystem functionality.
5. This is also a time when the idea of salutogenic, or health-promoting, environments is proposed (Antonovsky, 1987).

6. Environmental refugees were predicted to reach as many as 50 million by 2010 (Collins, Roberts, & Yano, 2005).
7. The PROCEED portion of Green and Kreuter's (1999) PRECEDE–PROCEED model of health promotion planning does this by recognizing the effect that policies have on the environment.
8. As a peculiar sidenote, there was a plan by a British fellow in the USA to build a city called Hygeia in present-day Kentucky decades before Richardson's book. The plan was never realized.
9. This work was originally published in 1898 as *To-Morrow: A Peaceful Path to Real Reform* and reissued in 1902 under its present title, *Garden Cities of To-Morrow*.
10. I was unable to find the original source of this often cited recommendation which appears to have originated around 1997.
11. This is not to imply that there was no such thing as urban planning done before this time. The Romans revolutionized the potential of cities with their plans. The physical remnants of Mayan cities and ancient cities in South Asia, as but two examples, were also undeniably planned. There is simply a lack of evidence of how they were done so. This is why I am careful to note the emergence of the field of urban planning as "modern" urban planning here.
12. It would be remiss to not also mention the decades of work on ecological models of health promotion done by Lawrence Green. Green, Richard, and Potvin (1996) present the evolution of ecological thinking in health promotion and its pervasiveness in many disciplines that relate to public health. This is explored in greater detail when covering the health behavior of physical activity in Chapter 6.
13. Parkes et al. (2003, Table 1) present a detailed chronology of the evolution and convergence of ecological thinking in the environmental and health sciences.
14. Hancock (1993) has also constructed a number of models (the Model of Human Development; the Model of Health and the Community Ecosystem) which squarely recognize the significance of the natural environment in an ecological model of health and the synergy between health and sustainable development.
15. Barton and Grant attribute Dahlgren and Whitehead (1991) with an antecedent to this figure. The original 1991 report with a slightly revised version of the antecedent figure can be found in Dahlgren and Whitehead

(2007a, p. 11). The slightly revised version of the figure can also be found in Dahlgren and Whitehead (2007b).

16. Northridge, Sclar, and Biswas (2003) provide a framework that includes the natural environment and highlights social determinants with much more detail.

17. Although not the focus of this book, there has also been an ecologic revolution in the role of the social environment on health behavior (McLeroy, Bibeau, Steckler, & Glanz, 1988). The social environment, rightfully so, continues to receive a great deal of attention in the health literature; the physical environment much less so.

References

Antonovsky, A. (1987). *Unraveling the mystery of health: How people manage stress and stay well*. Hoboken, NJ: Jossey-Bass.

Ashton, J. (1991). Sanitarian becomes ecologist: The new environmental health. *British Medical Journal, 302*(6770), 189–90.

Ashton, J., & Ubido, B. (1991). The healthy city and the ecologic idea. *Journal of the Society for the Social History of Medicine, 41*, 173–80.

Barton, H., & Grant, M. (2006). A health map for the local human habitat. *Journal of the Royal Society for the Promotion of Health, 126*(6), 252–3.

Berry, W. (1993). *Sex, economy, freedom & community: Eight essays*. New York, NY: Pantheon Books.

Bronfenbrenner, U. (1977). Toward an experimental ecology of human development. *American Psychologist, 32*, 513–31.

Bronfenbrenner, U. (1979). *The ecology of human development: Experiments by nature and design*. Cambridge, MA: Harvard University Press.

Bronfenbrenner, U. (1994). Ecological models of human development. In *International Encyclopedia of Education* (2nd ed., pp. 37–43). Oxford: Elsevier.

Brown, C., & Grant, M. (2005). Biodiversity & human health: What role for nature in healthy urban planning? *Built Environment, 31*(4), 326–38.

Collins, T., Roberts, I., & Yano, N. (2005). *As ranks of "environmental refugees" swell worldwide, calls grow for better definition, recognition, support*. Tokyo: United Nations University Institute for Environment and Human Security.

Coutts, C. (2010). Public health ecology. *Journal of Environmental Health, 72*(6), 53–5.

Coutts, C., Forkink, A., & Weiner, J. (2014). The portrayal of nature in the evolution of the ecological public health paradigm. *International Journal of Environmental Research and Public Health, 11*(1), 1005–19.

Dahlgren, G., & Whitehead, M. (1991). *Policies and strategies to promote social equity in health.* Copenhagen: WHO Regional Office for Europe.

Dahlgren, G., & Whitehead, M. (2007a). *Policies and strategies to promote social equity in health.* Stockholm: Institute for Future Studies.

Dahlgren, G., & Whitehead, M. (2007b). *Strategies for tackling social inequities in health: Levelling up part 2.* Copenhagen: WHO Regional Office for Europe.

Duhl, L. J. (1963). *The urban condition: People and policy in the metropolis.* New York, NY: Basic Books.

Duhl, L. J., & Sanchez, A. K. (1999). *Healthy cities and the city planning process: A background document on links between health and urban planning.* Copenhagen: WHO Regional Office for Europe.

Erickson, A. (2012). A brief history of the birth of urban planning. Retrieved from www.theatlanticcities.com/jobs-and-economy/2012/08/brief-history-birth-urban-planning/2365/

Evans, R., & Stoddart, G. (1990). Producing health, consuming health care. *Social Science & Medicine, 31*(12), 1347–63.

Fishman, R. (1982). *Urban utopias in the twentieth century: Ebenezer Howard, Frank Lloyd Wright, Le Corbusier.* Cambridge, MA: MIT Press.

Fishman, R. (1989). *Bourgeois utopias: The rise and fall of suburbia.* New York, NY: Basic Books.

Force, J., & Machlis, G. (1997). The human ecosystem part II: Social indicators in ecosystem management. *Society & Natural Resources: An International Journal, 10*(4), 369–82.

Glaser, M., Christie, P., Diele, K., Dsikowitzky, L., Ferse, S., Nordhaus, I., et al. (2012). Measuring and understanding sustainability-enhancing processes in tropical coastal and marine social–ecological systems. *Current Opinion in Environmental Sustainability, 4*(3), 300–8.

Green, L. W., & Kreuter, M. W. (1999). *Health promotion planning: An educational and ecological approach* (Vol. 3). Mountain View, CA: Mayfield.

Green, L. W., Richard, L., & Potvin, L. (1996). Ecological foundations of health promotion. *American Journal of Health Promotion, 10*(4), 270–81.

Hall, P. (2002). *Cities of tomorrow* (3rd ed.). Saffron Walden, UK: Blackwell.

Hancock, T. (1985). The mandala of health: A model of the human ecosystem. *Family & Community Health, 8*(1), 1–10.

Hancock, T. (1986). Lalonde and beyond: Looking back at "a new perspective on the health of Canadians." *Health Promotion, 1*(1), 93–100.

Hancock, T. (1993). Health, human development and the community ecosystem: Three ecological models. *Health Promotion International, 8*(1), 41–7.

Hancock, T., & Perkins, F. (1985). The mandala of health: A conceptual model and teaching tool. *Health Education, 24*, 8–10.

Hebbert, M. (1999). A city in good shape: Town planning and public health. *Town Planning Review, 70*(4), 433–53.

Heemskerk, M. (2003). Conceptual models as tools for communication across disciplines. *Conservation Ecology, 7*(3), online.

Hongxun, Y. (1982). *Classical gardens of China: History and design techniques.* New York, NY: Van Nostrand Reinhold.

Howard, E. (1965). *Garden cities of to-morrow.* Cambridge, MA: MIT Press.

Howard, E. (2009). *Garden cities of to-morrow (Illustrated Edition).* Gloucester, UK: Dodo Press.

Johnson, S. (2007). *The ghost map: The story of London's most terrifying epidemic—and how it changed science, cities, and the modern world.* New York, NY: Penguin.

Kickbusch, I. (1989). Approaches to an ecological base for public health. *Health Promotion International, 4*(4), 265–8.

Lalonde, M. (1974). *A new perspective on the health of Canadians.* Ottawa: Ministry of Supply and Services Canada.

LeGates, R. T., & Stout, F. (2011). *The city reader.* Abingdon: Routledge.

Machlis, G., Force, J., & Burch Jr, W. (1997). The human ecosystem part I: The human ecosystem as an organizing concept in ecosystem management. *Society & Natural Resources: An International Journal, 10*(4), 347–67.

Maller, C., Townsend, M., Pryor, A., Brown, P., & St Leger, L. (2006). Healthy nature healthy people: "Contact with nature" as an upstream health promotion intervention for populations. *Health Promotion International, 21*(1), 45–54.

Martensen, R. (2009). Landscape designers, doctors, and the making of healthy urban spaces in 19th century America. In L. Campbell & A. Wiesen (Eds.), *Restorative commons: Creating health and well-being through urban landscapes* (pp. 26–37). Newtown Square, PA: US Department of Agriculture, Forest Service, Northern Research Station.

Maslow, A. H. (1966). *The psychology of science.* Richmond, CA: Maurice Bassett.

McCally, M. (2002). Environment, health, and risk. In M. McCally (Ed.), *Life support: The environment and human health* (pp. 1–14). Cambridge, MA: MIT Press.

McKeown, T. (1971). A historical appraisal of the medical task. In G. McLachlan & T. McKeown (Eds.), *Medical history and medical care* (pp. 1–23). London: Oxford University Press.

McKeown, T. (1979). *The role of medicine: Dream, mirage or nemesis?* (2nd ed.). Oxford: Blackwell.

McLeroy, K., Bibeau, D., Steckler, A., & Glanz, K. (1988). An ecological perspective on health promotion programs. *Health Education Quarterly, 15*(4), 351–77.

McMichael, A. J., Campbell-Lendrum, D., Corvalan, C., Ebi, K., Githeko, A., Scheraga, J., & Woodward, A. (2003). *Climate change and human health: Risks and responses*. Geneva: WHO.

Millennium Ecosystem Assessment. (2005). *Ecosystems and human well-being: Health synthesis*. Geneva: WHO.

Morris, G. P., Beck, S. A., Hanlon, P., & Robertson, R. (2006). Getting strategic about the environment and health. *Public Health, 120*(10), 889–903.

Muir, J. (1912). *The Yosemite*. New York, NY: Century Company.

Myers, S. S., & Patz, J. A. (2009). Emerging threats to human health from global environmental change. *Annual Review of Environment and Resources, 34*(1), 223–52.

Myers, S. S., Gaffikin, L., Golden, C. D., Ostfeld, R. S., Redford, K. H., Ricketts, T. H., et al. (2013). Human health impacts of ecosystem alteration. *Proceedings of the National Academy of Sciences, 110*(47), 18753–60.

Niemelä, J. (1999). Is there a need for a theory of urban ecology? *Urban Ecosystems, 3*(1996), 57–65.

Niemelä, J., Saarela, S.-R., Söderman, T., Kopperoinen, L., Yli-Pelkonen, V., Väre, S., & Kotze, D. J. (2010). Using the ecosystem services approach for better planning and conservation of urban green spaces: A Finland case study. *Biodiversity and Conservation, 19*(11), 3225–43.

Northridge, M. E., Sclar, E. D., & Biswas, P. (2003). Sorting out the connections between the built environment and health: A conceptual framework for navigating pathways and planning healthy cities. *Journal of Urban Health, 80*(4), 556–68.

Olival, K. J., Hoguet, R. L., & Daszak, P. (2013). Linking the historical roots of environmental conservation with human and wildlife health. *EcoHealth, 10*(3), 224–7.

Parkes, M., Panelli, R., & Weinstein, P. (2003). Converging paradigms for environmental health theory and practice. *Environmental Health Perspectives, 111*(5), 669–75.

Pavao-Zuckerman, M. (2000). The conceptual utility of models in human ecology. *Journal of Ecological Anthropology, 4*, 31–56.

Pickett, S., Cadenasso, M., & Grove, J. (2001). Urban ecological systems: Linking terrestrial ecological, physical, and socioeconomic components of metropolitan areas. *Annual Review of Ecology and Systematics, 32*, 127–57.

Rapport, D., Gaudet, C., Constanza, R., Epstein, P., & Levins, R. (Eds.). (1998). *Ecosystem health: Principles and practice.* Oxford: Wiley-Blackwell.

Rauch, J. (1869). *Public parks: Their effects upon the moral, physical and sanitary condition of the inhabitants of large cities; with special reference to the city of Chicago.* Chicago, IL: SC Griggs.

Richardson, B. W. (1876). *Hygeia: A city of health.* London: Macmillan.

Schulz, A., & Northridge, M. E. (2004). Social determinants of health: Implications for environmental health promotion. *Health Education & Behavior, 31*(4), 455–71.

Smith, K. R., & Ezzati, M. (2005). How environmental health risks change with development: The epidemiologic and environmental risk transitions revisited. *Annual Review of Environment and Resources, 30*(1), 291–333.

Town and Country Planning Association. (2012a). *Creating Garden Cities and suburbs today: Policies, practices, partnerships and model approaches.* London: Town and Country Planning Association.

Town and Country Planning Association. (2012b). *Reuniting health with planning: Healthier homes, healthier communities.* London: Town and Country Planning Association.

VanLeeuwen, J. A., Waltner-Toews, D., Abernathy, T., & Smit, B. (1999). Evolving models of human health toward an ecosystem context. *Ecosystem Health, 5*(3), 204–19.

Walker, S. E., & Duffield, B. S. (1983). Urban parks and open spaces—an overview. *Landscape Research, 8*(2), 2–12.

Ward Thompson, C. (1998). Historic American parks and contemporary needs. *Landscape Journal, 17*(1), 1–25.

Ward Thompson, C. (2011). Linking landscape and health: The recurring theme. *Landscape and Urban Planning, 99*(3–4), 187–95.

Warpole, K. (2007). The health of the people is the highest law: Public health, public policy and green space. In C. Ward Thompson & P. Travlou (Eds.), *Open space: People space* (pp. 11–22). Abingdon: Taylor & Francis.

Watts, J., & Borger, J. (2011). Tsunami pushes Japan from major aid donor to leading recipient. *The Guardian*, May 11.

WHO. (n.d.). Ottawa charter for health promotion. In *First International Conference on Health Promotion*. Ottawa: WHO.

Williams, H. (1944). Public health and urban planning. In B. Hovde, T. MacDonald, G. McLaughlin, & H. Williams (Eds.), *Report of the Urban Planning Conferences at Evergreen House, 1943, under the auspices of Johns Hopkins University* (pp. 115–21). Baltimore, MD: Johns Hopkins University Press.

PART II

THE ECOSYSTEM SERVICES SUPPORTED BY GREEN INFRASTRUCTURE

Chapter Three

Essential ecosystem services

The dependency of human health on basic ecosystem services is the most funda-mental, most catastrophic if threatened, and often most overlooked function of GI presented in this book. Ecological models of health reflect the fundamental importance of ecosystems to health and the role of the natural environment in supporting them, but it has only been within the past decade that public health has begun to seriously confront the threat to health posed by degraded ecosystems. This is despite a much longer history of the recognized importance of the human place within ecosystems, our dependence upon the services they deliver, and the concurrent "quiet crisis" of ecological degradation (Millennium Ecosystem Assessment, 2005; Sanesi, Gallis, & Kasperidus, 2011). A steady awakening is reflected most notably in the World Health Organization report, *Ecosystems and Human Well-Being: Health Synthesis*, which illustrates the complex way that health is linked to the integrity of ecosystems (Millennium Ecosystem Assessment, 2005). In the following sections, we review how GI is vital to ecosystem services as basic as water, air, and food, and to maintain the biologic reservoir of medicinal compounds used to treat disease.

Water

Green infrastructure plays a vital role in the continued provision and control of the quantity and quality of *the* most essential of life-supporting elements: water. The role GI plays in regulating water quantity stems foremost from its role in the hydrological cycle, but GI is also important to water quantity because it facilitates the recharge of groundwater stores and aids in controlling surface runoff volumes. Green infrastructure supports water quality through its ability to filter pollutants from rainfall and pollutants that are collected in surface runoff.

Green infrastructure, and forests in particular, play a key role in the hydrological cycle as GI facilitates the infiltration and storage of water in soils and the release of water back into the air through transpiration (Tyrväinen, Pauleit, Seeland, & de Vries, 2005) (Figure 3.1). Transpiration is the process by which water is drawn out of the soil and released into the air through the process of plants "breathing" (US Geological Service, 2013). Rain brings this moisture back to earth to replenish surface water sources, and its migration through soils leads to its accumulation in groundwater reservoirs. Surface water also wells up from groundwater sources. The contribution of GI to rainfall is also self-serving to GI as rainfall is, of course,

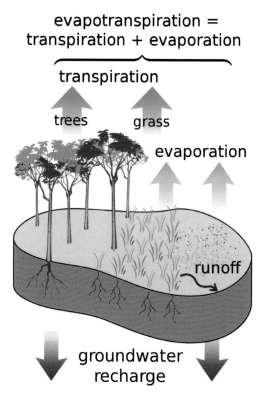

Figure 3.1: Evapotranspiration
Source: Mwtoews.

essential to sustain the GI that transports water through the hydrological cycle. While this may seem an elementary review of basic science, it is doubtful that many people regularly reflect on the role of GI in this process and regard actions that degrade local and global GI as threats to the provision of water. Protecting the GI that supports the hydrological cycle could be considered public health promotion at its most fundamental level.

Over a billion persons on earth depend on groundwater reservoirs as their primary source of drinking water (Table 3.1) (Sampat, 2000). Groundwater is almost the exclusive source of water for inhabitants of some of the largest cities on earth (e.g. Mexico City, Dhaka). It is also the exclusive source of water in many rural areas. In the USA, with its extensive yet aging infrastructure, 95 percent of the population depends on groundwater for drinking and household needs. Groundwater is also essential not only to drinking and household needs related to health (e.g. personal hygiene), but also to producing another commodity essential to life: food. Groundwater is used extensively in the irrigation of crops. Aquifers supply water to more than half of the irrigated land in India and to nearly half of the irrigated land in the USA (Sampat, 2000, p. 11). For coastal communities ("water, water, everywhere, nor any drop to drink," Samuel Taylor Coleridge, *The Rime of the Ancient Mariner*) desalination would appear to provide an inexhaustible source of water and a remedy for depleted aquifers, but, except for a few places on earth, this process has yet to be developed to a level where it is viable. Until it is, groundwater will remain an essential source of water to those who have the infrastructure necessary to extract it. For all, whether dependent on groundwater or surface water, as GI is diminished, so too will be the quantity of water available for human use.

Table 3.1: Groundwater as share of drinking water

Region	Share of drinking water (%)	People served (millions)
Asia–Pacific	32	1,000 to 1,200
Europe	75	200 to 500
Latin America	29	150
United States	51	135
Australia	15	3
Africa	n/a	n/a
Total global		1,500 to 2,000

Source: Adapted from Sampat (2000).

Green infrastructure is also important to regulating the quality of water. The many forms of GI, from forests to green roofs on buildings, act as natural filters of the pollutants that can accumulate in surface and groundwater. Many of these pollutants are toxic to human beings and can accumulate in the body through the direct consumption of polluted water and indirectly through the consumption of animals and plants that consume polluted water. Rainwater runoff carried unchecked over impervious surfaces,[1] such as asphalt and concrete, and into surface water bodies carries with it the pollutants deposited on these surfaces. This subsequently increases the levels of non-point source pollution in surface water and groundwater (Arnold & Gibbons, 1996). These pollutants include metals, organic compounds, and excessive nutrients. Most of the metals and organic compounds stem from vehicular combustion and operation (Lukes & Kloss, 2008). Some of the pollutants that accumulate in surface runoff are atmospheric pollutants released by humans and then brought back to the earth in rainfall.

Green infrastructure has the ability to filter pollutants that fall with the rain and then accumulate in surface runoff. "As [runoff] percolates through the soil, the water is stripped of many [pollutants], both by being taken up by plants and microbes and by coming into contact with chemically reactive sites on clay and on organic matter to which such compounds bind" (Melillo & Sala, 2008, p. 84). In coastal communities where untreated runoff flows into the sea and onto public beaches, it has been found that surfers on these beaches experienced twice as many health problems as those on beaches not exposed to urban runoff (Dwight, Baker, Semenza, & Olson, 2004).

Green infrastructure is also important to controlling runoff volumes. It has been shown that "runoff volumes in vegetated areas are typically between 10–20% of the average annual rainfall. In urban areas, where surfaces are highly impervious, typical runoff volumes are 60–70% of the average annual rainfall" (Minnesota Pollution Control Agency, 2000, pp. 4.20–1). As is witnessed with increasing frequency in many parts of the world, flooding has dire consequences for health due to injuries, infectious disease transmission, and the disruption of physical and social systems.

Figure 3.2 is a stylized representation of a GI strategy used to control urban runoff at the urban block level. Green roofs, vegetated road medians, trees, and permeable paving capture rainwater that would otherwise run unchecked over impermeable surfaces collecting pollutants and running in higher volumes. This is truly putting the "infrastructure" in green infrastructure because GI here is serving

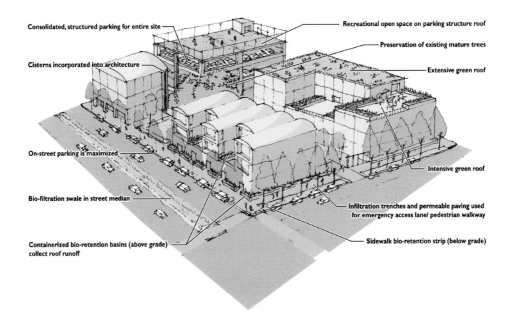

Figure 3.2: Block level green infrastructure

Source: Clark Wilson, Community Design + Architecture for City of Emeryville, California.

a function typically associated with gray infrastructure. The GI system here at the block level could be considered a microcosm of a larger GI system. The GI system in Figure 3.2 should connect to GI in other blocks, neighborhoods, and regions to form the ecological GI complex necessary to regulate water quantity and quality and support health. This GI complex is also critical to ensuring the quality of another essential element of life: air.

Air

Towering smokestacks releasing a visible blanket of poison into the air are remnants of a bygone industrial era in some parts of the world but are still very much a daily threat to health in others. Smokestack emissions are an obvious culprit in diminished air quality, but other more easily disguised threats to air quality, such as the global increase in automobile use, are also, if not a more, serious threat. Progress has been made in recognizing that the control of air pollution is a global issue, but collaborative regional and global efforts have been painfully slow in reducing air pollution at its various sources. Until local, national, and global efforts are successful in reducing pollution at its source, it is critical to local, national, and global health to implement adaptive strategies to remediate the pollutants currently being released and diffused in the atmosphere. Green infrastructure conservation is one such strategy. Green infrastructure, and trees in particular, can reduce air pollution related morbidity through its natural ability to capture diffused air pollution. Green infrastructure may also help mitigate the release of air pollution by supporting non-motorized forms of transportation.[2]

David Nowak has been one of the most prolific voices and contributors to the science of the ecosystem services of trees. He has developed a clever acronym, **TREE**, to summarize the ecosystem services of trees including their direct and indirect air cleansing functions (Nowak, 1995).

Temperature reduction and other microclimatic effects
Removal of air pollutants
Emission of volatile organic compounds and tree maintenance emissions
Energy effects on buildings.

Focusing on the potential threats to health trees can aid in ameliorating, trees can contribute to reducing ambient heat (**T**) and removing pollutants directly from the air (**R**). Trees can also reduce the energy needs of buildings and associated emissions (second **E**). Granted, it is not all rosy. Trees themselves emit volatile organic compounds (VOCs) that react with other pollutants that threaten health, and activities to maintain GI can also result in the emission of air pollutants (first **E**). The fossil fuels combusted while mowing, pruning, and leaf blowing in parks may outweigh the benefits these spaces provide in carbon sequestration (Oliver-Solà & Núñez, 2007).

Discussed in this section is the removal of air pollutants (**R**) by trees and other forms of GI. Also covered here is the emission of VOCs by trees (first **E**). The health benefits of the temperature reduction properties of GI (**T**) and the energy effects (second **E**) related to carbon capture and climate change are covered in Chapter 4 (the "Green Infrastructure and Weather and Climatic Events: Adaptation and Secondary Prevention" and "Green Infrastructure and Carbon Sequestration: Mitigation and Primary Prevention" sections, respectively). As with most of the relationships between GI and health, the overlap in the functions and benefits of GI makes them difficult to separate. For example, the amount of fossil fuel energy a building demands determines the level of air pollution it emits. The level of air pollution it emits influences its contribution to raising ambient heat which then influences the levels of energy demanded to counteract ambient heat. Green infrastructure intervenes at all points of this cycle. Green infrastructure can capture air pollution, reduce ambient heat, and reduce the energy consumption that leads to air pollution and more ambient heat, but not just any GI will do. The design of GI, with regard to wind, shade, the species chosen in design, and many other factors can optimize the beneficial health effects of GI in improving air quality (Domm et al., 2008; Hartig, Mitchell, de Vries, & Frumkin, 2014; Nowak, Crane, & Stevens, 2006).

One might be questioning: "Since we are talking about GI and air, what about the essential role of plants in producing the oxygen we breathe?" A fair question. Photosynthesis is responsible for 98 percent of the oxygen in the atmosphere with ultraviolet radiation breaking apart water molecules accounting for the other 2 percent. Half, and possibly a bit more, of the oxygen in the atmosphere is produced by phytoplankton in the oceans. The other half of the oxygen we breathe is produced by terrestrial plant life or GI. Nicholson-Lord (2003), in summarizing the many benefits of GI in cities, claims that many cities have as little as 10–12 percent oxygen in the atmosphere as compared with the more normal occurrence of about 21 percent, and a large, mature tree can produce enough oxygen[3] for 10 people's daily needs. Nowak, Hoehn, and Crane (2007) calculated that it takes 30 trees to supply one adult's annual oxygen needs. This brings the number of persons daily needs met to 12 (30 trees/365 days = 0.082 trees needed to meet 1 person's daily needs, 1 day/.082 trees = 12.2 persons daily needs met). Other sources quote the estimate of four persons daily oxygen needs being met by a mature tree, although it is not clear what the source of this estimate is or how it was derived (Environment Canada, 2012). In the USA alone, the amount of oxygen that urban trees produce is enough "to offset the annual oxygen consumption of approximately two-thirds of the population," but we are cautioned that it is the other benefits of forests that "are more critical to environmental quality and human health than oxygen production by urban trees" (Nowak et al., 2007, p. 220). There are number of reasons why oxygen production is likely not the greatest service terrestrial GI provides to local and global health. This has to do with the complexity of determining the "normal" levels of oxygen humans need (think of those who live at high altitudes), the fact that oxygen diffuses rapidly in the air due to a wide range of climatic and built environment conditions, and the reality that any diminishment of terrestrial GI is not likely to pose any near-term threat to the large supply of oxygen already in the atmosphere (half of all oxygen is produced by aquatic vegetation). Maintaining local and global supplies of oxygen is not a major or immediate benefit of terrestrial GI. What is a significant health benefit is the ability of GI to filter air pollutants that, even in small concentrations, can negatively impact environmental quality and human health (Nowak et al., 2006).

Selected forms of vegetation, and again most notably trees, have the capacity to capture both gaseous and particulate airborne pollutants (Bealey et al., 2007; Beckett, Freer-Smith, & Taylor, 1998). Gases are removed from the air by the uptake by leaf stomata (pores on leaves), absorption through leaf surfaces, and adsorption (or adherence) to plant surfaces. Particulate matter removal occurs through deposition on leaves and other plant surfaces. Particulate matter is held on these surfaces by either being stuck on impact or through an adsorptive chemical process (Hedin, 2000).[4]

Trees remove tons of air pollution annually. In the southern US cities of Houston and Atlanta, both with similar tree coverage, annual removal of particulates by trees was 4.7 and 3.2 tons per square mile, respectively (Wolf, 2004). In fact, urban trees alone were estimated to be responsible for the removal of 711,000 metric tons (1,567,486,684 pounds) of pollutants annually, a $3.8 billion value (Nowak et al., 2006).[5] Intertwined in the value of trees could be perceived health benefits, but this externality is not explicit and can only be assumed. A more recent study explicitly focused on health benefits found the value of the avoidance of human mortality and acute respiratory symptoms attributed to the uptake of air pollution by trees and forests in the USA in 2010 to be $6.8 billion (Nowak, Hirabayashi, Bodine, & Greenfield, 2014). In other cities struggling with air pollution issues such as Beijing, it was estimated that trees removed 1261.4 tons of pollutants from the air in one year (Yang, McBride, Zhou, & Sun, 2005). Urban trees have also been found to be important to the ecosystem service of air pollution abatement in other rapidly expanding Chinese cities (Jim & Chen, 2008).

A handful of studies have tied the ability of trees to capture pollutants to morbidity. In New York City neighborhoods, an association was found between tree density and a lower prevalence of early childhood asthma (Lovasi, Quinn, Neckerman, Perzanowski, & Rundle, 2008), but Lovasi et al. (2013) have since determined the validity of these findings may have been compromised by the scale of their analysis and the limited tree type included in their analysis. Employing a natural experiment study design, Donovan et al. (2013) examined the natural destruction of trees across massive swaths of the eastern USA by the emerald ash borer to discern if the loss of trees had any effect on cardiovascular and lower respiratory mortality. They were able to reveal a strong relationship between many more thousands of deaths from these two causes and the destruction of trees. A number of ecosystem services presented in this book could be potential mechanisms that help explain this relationship, but it is suspected that air pollution abatement in particular is likely playing an important role especially when considering the outcome of lower respiratory mortality.

The pollutants trees remove are: ozone (O_3), nitrogen oxides (NO_x), sulfur dioxide (SO_2), particulate matter with a diameter of 10 micrometers or less (PM_{10}), and carbon monoxide (CO).[6] Some of these airborne pollutants are direct threats to health, others are important because they react with other pollutants such as volatile organic compounds (VOCs). Stopping far short of a lesson in atmospheric chemistry that I am ill-equipped to deliver, let us briefly examine each one of these pollutants that GI has the ability to abate.

Ozone forms in the troposphere and at ground level in reactions with NO_x and VOCs.[7, 8] Concentrations of ozone are higher in sunny areas due to the photochemical reaction between NO_x and VOCs. Ozone is not a serious threat to health in low concentrations, the typical range of ambient ozone being 60–120 ppb, but in higher concentrations ozone can inflame airways and reduce lung function (US EPA, 2013b). It is important to note that individual reactions to the same level of ozone can vary widely from no reaction to severe respiratory distress. In the least severe cases, ozone may cause coughing and throat irritation. In the worst cases, generally among those who suffer from decreased lung function from conditions such as asthma, exposure to ozone can cause shortness of breath and even death. Trees help to reduce ozone by capturing ozone molecules directly from the air with a small, yet measurable, effect (Nowak, Civerolo, Trivikrama Rao, Luley, & Crane, 2000), but trees also reduce ozone due to their ability to capture VOCs.

The greatest direct threats to health from VOCs come from their high levels in indoor environments.[9] Out-of-doors, burning fossil fuels and plant matter releases the VOCs that contribute to ozone formation. In the USA, approximately half of human VOC emissions is attributed to on-road and non-road engines and vehicles (US EPA, 2000). Likely surprising to most is that trees themselves are responsible for the release of VOCs, accounting for approximately two-thirds of all VOCs currently in the air (Naranjo, 2011). Due to this, trees could lead to actual increases in ozone (Du, Kang, Lei, & Chen, 2007), but this only occurs when concentrations of accompanying NO_x pollutants are high. It is when NO_x is present in low concentrations that VOCs can reduce ozone. The science of VOC emissions by trees and vegetation is evolving, and recent studies suggest that human-produced pollution and increased temperatures spur trees and vegetation to increase their level of VOC emission (Carlton & Pinder, 2010). Likewise, the role of deciduous vegetation in removing oxygenated VOCs (oVOCs) from the atmosphere is evolving, with some models suggesting that increases in airborne pollutants actually spur deciduous vegetation to increase their uptake of oVOCs at levels considerably higher than previously thought (Karl, Harley, Emmons, & Thornton, 2010). Volatile organic compounds are detrimental to health due to their role in ozone formation but also because they create secondary organic aerosols (SOAs). SOAs are important to the hydrological cycle due their role in seeding clouds and to their role in holding atmospheric heat, but they are also a harmful form of particulate matter (Perraud et al., 2012).

NO_x is similar to VOCs in that it is only a direct threat to health indoors or in industrial settings where high concentrations could lead to serious respiratory distress (National Library of Medicine, 2013). It is also similar to VOCs in that it is most pertinent to health due to its role in ozone and particulate matter formation. Automobiles are a common source of NO_x, visible as the brown smog seen over many cities when it reacts with VOCs and sunlight. In this process, suspended particulate matter is also created.

Sulfur dioxide is another gaseous pollutant that, like VOCs and NO_x, is a risk to most people due to its role in forming airborne particulates. Trees have the ability to filter SO_2 from the air before it forms harmful particulates. In the Chicago area alone, trees are responsible for removing 3.9 tons of SO_2 daily (McPherson, Nowak, & Rowntree, 1994). While this may seem like a large absolute amount (it is), this translates into only a maximum hourly reduction of 1.3 percent of total SO_2. In areas of Chicago that were 100 percent forested, the maximum hourly reduction of SO_2 was five times that. These measurable effects in reducing SO_2, by far the greatest source of which is fossil fuel combustion (US EPA, 2013a), are an important service of trees. Reducing levels of SO_2 prevents it from reacting with other atmospheric compounds to form the particulates that penetrate deeply into the lungs (US EPA, 2015).

Particulate matter with a diameter of 10 micrometers or less, or PM_{10}, poses a risk to human health because these particulates are small enough to reach the lower respiratory tract and cause respiratory distress, lung tissue damage, cancer, and premature death (US EPA, 1995). Data from urban air quality monitoring suggest that particulates may represent a greater threat to human health than any other air pollutant (Beckett, Freer-Smith, & Taylor, 2000, p. 209). At the same time, there is promising evidence that trees can aid in the removal of these pollutants. Trees in Santiago, Chile, have been estimated to remove as much as $7.5g/m^2$ (67 lbs per acre) annually (Escobedo & Nowak, 2009). In Bangladesh, the canopy density of roadside vegetation was significantly associated with decreased levels of total suspended particulates, especially in summer months (Islam et al., 2012). The significance of tree density arises again with findings of decreased PM_{10} concentrations in the interior of a broadleaf evergreen urban forest (Cavanagh, Zawar-Reza, & Wilson, 2009). But not all trees are the same. The rate of uptake of particles varies between species, with trees with more rough leaf surfaces performing the best as filters of particulate matter. While mature trees have been found in most studies to be the most effective form of GI for reducing air pollution, there is concurrent evidence "that younger, smaller trees of the same species are also highly effective at removing pollutant particles due to their greater foliage densities" (Beckett et al., 2000, p. 209). Trees can also aid in the removal of the much smaller diameter particulates ($PM_{2.5}$) that can find their way very deep into lung tissue and increase the risk of lung cancer and cardiopulmonary mortality (Nowak, Hirabayashi, Bodine, & Hoehn, 2013; Pope et al., 2002). The influence of $PM_{2.5}$ on mortality from a number of causes has been shown to be higher than previously observed for PM_{10} (Zanobetti & Schwartz, 2009). Taken together, this evidence can guide the design choices that can optimize trees as particulate matter removers. The choice of species, density of planting, and age of trees that compose GI will have an influence on the optimization of benefits from reducing PM_{10} and $PM_{2.5}$ and the gases that contribute to their formation.

Plants also capture carbon monoxide and carbon dioxide (excluded from the VOC designation along with a handful of other organic compounds) that, in high concentrations, can produce headaches, dizziness, nausea, and death from asphyxiation. These gases rarely occur in concentrations out-of-doors that are harmful to health. Green infrastructure is much more important at reducing the health risks posed by these gases through the ability of plants to sequester the carbon from these molecules in their biomass. This carbon sequestration ecosystem service is discussed in Chapter 4, "The Challenge of Climate Change."

Greenspace also has some benefit to not only the remediation of gaseous and particulate pollutants but also the prevention of their release. Greenways are one form of GI that can provide the environmental support to substitute motorized travel (the culprit in many airborne pollutants) with walking and biking. When the health effects of air pollution have been included among all the external costs of motorized transportation, it has been found that "the most important external costs are health costs due to cardiovascular and respiratory diseases caused by air pollutants" (van Essen et al., 2011, p. 33). Greenway users are keen at making the connection between using transportation other than a car and decreasing levels of pollution (Shafer, Lee, & Turner, 2000). The co-benefit to health, of course, is the physical activity gained through non-motorized forms of transportation and recreation. The great need for a better accounting of these co-benefits, especially as it relates to air pollution, has been raised in the scientific literature (Jack & Kinney, 2010).

While there is mounting empirical evidence to support Frederick Law Olmsted's century-old claim that parks and trees are the "lungs of the city," GI is not a stand-alone solution for cleansing the air for an expanding global population. Green infrastructure is part of the solution, but taking into account the current rate of emissions, the sum of pollutants already in the atmosphere, and the potential air quality trade-offs of GI implementation (e.g. VOCs), GI is likely to play only a minor role in reducing air pollution (Pataki et al., 2011; Whitlow et al., 2014). Air pollution not only has a direct effect on human health, but it also threatens the GI and other forms of life that are necessary for the production of other ecosystem services. We have now covered the critical role of GI in the ecosystem services related to water and air, but we are also dependent on GI for another basic element of life: the food we eat.

Food

The production of food depends on three ecosystem processes in which GI plays a vital role. These are primary production, nutrient cycling, and pollination. Primary production is "the synthesis and storage of organic molecules during the growth and reproduction of photosynthetic organisms" (Kling, 2008). Photosynthetic organisms—trees, plants, and some bacteria—feed themselves by capturing energy from the sun. Humans use 40 percent of the earth's net primary production of plant material for food, fiber, and lumber (Vitousek, Ehrlich, Ehrlich, & Matson, 1986). Other estimates are closer to 25 percent (Krausmann et al., 2013), which is still stunning. Of all the energy that is captured by all the plants on earth, humans consume or co-opt a quarter to almost half of it. Some of the plant material from primary production is used by humans for building materials, burning for fuel, and for clothing; however, I am focusing here on how there would be no food without plants and GI. Now, if you are thinking, "I'm not a vegetarian, so what does it matter?" then you may be in store for a small surprise (or maybe just a reminder). The autotrophs, such as plants, that feed themselves through photosynthesis also feed *all* other organisms on earth ... eventually. As a heterotroph, humans can only obtain energy from feeding on other organisms. These organisms are autotrophs that capture energy from the sun or other heterotrophs that either capture energy from autotrophs or feed on other heterotrophs that somewhere down the food chain fed on autotrophs. Without the primary production of plant material, there would be no food from either plant or animal sources and human life would cease. Green infrastructure is therefore essential to maintaining the primary production that is currently being lost through land development, selective overharvesting, and deforestation. At the very least, particularly for those who subsist on locally grown seasonal crops, we will continue to witness the many public health maladies caused by malnutrition. At worst, the primary production that is lost due to diminished GI will lead to food shortages and starvation.

Patrick Geddes, an eclectic and influential thinker who was also an early urban planning theorist, is noted as saying in his farewell lecture at Dundee University "How many people think twice about a leaf? Yet the leaf is the chief product and phenomenon of Life: this is a green world, with animals comparatively few and small, and all dependent upon the leaves. By leaves we live" (Defries, 1927). The loss of the leaves is not just a direct threat to our food sources, but it is also a threat to the ability of soil to support plants. The loss of the leaves, GI, and primary production degrades the ability of plants to release nutrients back into the soil. The production and eventual death and decomposition of plants (and the heterotrophs that feed on them) are essential to the supporting ecosystem service of nutrient cycling. The flow of the macronutrients of carbon, hydrogen, oxygen, nitrogen, phosphorus, and sulfur through ecosystems is essential to produce food and to build living tissue. Green infrastructure supports the cycling of nutrients through ecosystems and also provides the habitat for the organisms that pollinate crops.

Pollination is another supporting ecosystem service needed for food production and dependent on GI. The majority of the world's crops consumed by humans are completely to at least moderately dependent on animal-mediated pollination, and a diminished landscape and biodiversity jeopardizes the ability of pollinators to do their job (Klein et al., 2007). Green infrastructure provides the habitat necessary for the bees, moths, butterflies, beetles, and bats that carry pollen from male to female plants that then bear fruits and vegetables that humans consume (or that are fed to other animals that humans later consume). It also provides the habitat for the biological control agents that prey on crop pests (Hillel & Rosenzweig, 2008).[10] Even modern industrial agriculture depends heavily on pollinators to produce foods such as cucumber, pear, apple, cherry, watermelon, broccoli, blueberry, almond, and many others. The annual economic value of the pollination service that bees alone provide has been estimated at $14.6 to $40 billion in the USA (Morse & Calderone, 2000; Pimentel et al., 1997). Other plants such as soybean and strawberry do not need insect pollinators, but yields are improved with them. Declines in certain types of pollinators providing this essential ecosystem service have been reported on every continent except Antarctica (Kearns, Inouye, & Waser, 1998), and it is strongly suspected that reduced habitat and biodiversity has played a role in these declines.

Crops that depend partially or fully on animal pollinators provide a large portion of essential dietary nutrients such as vitamins C and A, assorted antioxidants, calcium, fluoride, and folic acid (Eilers, Kremen, Smith Greenleaf, Garber, & Klein, 2011). And just in case there was any misconception that the diseases associated with malnutrition were just problems for those who practice subsistence style farming, the degradation to natural systems and GI can also increase heart disease, a leading cause of death in the developed world. There are two pathways by which this can occur. First, the loss of habitat of animal pollinators reduces the yields of crops that contain the nutrients (such as folic acid) that protect against chronic conditions such as heart disease. Second, as Myers et al. (2013) point out, increased risk of coronary heart disease may also occur from the substitution of dietary protein with dietary carbohydrate (Appel et al., 2005). This is more likely to occur as the protein levels in plant foods consumed by humans are reduced due to increased CO_2 concentrations (Taub, Miller, & Allen, 2008). Green infrastructure can reduce the CO_2 concentrations that are reducing the protein content of our food.

The Ecological Transition (Figure 2.12) reminds us that the wealthy will be able to compensate for the nutritional deficiencies of food caused by diminished GI, but the majority of people on earth will not have this luxury. It is in the developing nations of the world "that dwindling populations of marine and terrestrial wildlife may represent a nutritional crisis" as people in these contexts "cannot readily replace these foods with domesticated species or fortified foods" (Myers et al., 2013, p. 18754). In Madagascar, for example, it was found that if the population that depends on terrestrial wildlife as a protein source was unable to harvest it for consumption, children would experience a 29 percent higher risk of iron deficiency—the most prevalent nutritional deficiency worldwide (Golden, Fernald, Brashares, Rasolofoniaina, & Kremen, 2011). This condition has serious consequences on cognitive functioning and development, physical functions, and emotional regulation.

Furthermore, there is no distinction between medicine and healthy food in much of the developing world (Karjalainen, Sarjala, & Raitio, 2010, p. 4), and the rediscovery of this relationship is apparent in the rise of community gardens in many countries (Tranel & Handlin, 2006). Community gardens are part of a community's GI, and these forums for food production bring with them a host of other health benefits (Figures 3.3 and 3.4). People who participate in community

Figure 3.3: Davie Village Community Garden, Vancouver, Canada
Source: Daryl Mitchell.

Figure 3.4: Clementi Community Farm, Singapore
Source: Jnzl's Public Domain.

gardening not only report a desire for the higher quality food produced from them but also improved physical and mental health and social interactions (Armstrong, 2000). There is some evidence to suggest that all of these desires for improved health are being realized. The therapeutic benefits of community gardening (or therapeutic horticulture) stem from the physical activity gained through gardening, the mental health benefits of the affiliation with nature, and the social interactions inherent in this public activity (Aldridge & Sempik, 2002). The social interaction of communal gardening has been found to be important for older persons and particularly for the overall psychological well-being of those in long-term care facilities (Barnicle & Midden, 2003; Milligan, Gatrell, & Bingley, 2004). Community gardens have also been shown to bring a sense of well-being to diverse groups and to promote the bridging of interactions between these groups (Shinew, Glover, & Parry, 2004). The powerful communal and nutritional benefits of gardening are nowhere more evident than in "guerilla gardener" Ron Finley's inspirational message to take your neighborhood and your health back by just going outside to "plant some shit" (TED, 2013).

So, it is not just food itself that can be viewed as medicine but also the activities associated with growing healthy food that can provide therapeutic and prophylactic health benefits. Of course, using healthy food as a nutraceutical (food or food products that have health benefits) cannot prevent all ills. When GI's support of healthy food is not enough, GI also acts as an essential reservoir of the pharmaceuticals used to treat disease.

Medicine

The loss of GI and biodiversity has implications for biomedicine as well as public health. As the diversity of life on earth is diminished through the loss of GI, so too is the source of many pharmaceuticals currently in use and the unmeasured potential of yet to be discovered medicines from terrestrial and marine plants, animals, and microbes. Green infrastructure provides a rich reserve of compounds that can be utilized in pharmaceuticals (Karjalainen et al., 2010). "Thirty percent of the drugs sold worldwide contain compounds derived from plant material" (United Nations, 2004, p. 1), and at least half of all prescribed drugs in the USA

Figure 3.5: 'Iliahi or Forest Sandalwood, Hawaii

Source: David Eickhoff.

Note: The leaves of 'Iliahi can be used as a shampoo for dandruff and head lice. Drinking powdered material from this plant was used to treat "sores of long duration" on male and female sex organs.

come either directly from natural sources or are derived from natural sources (Grifo, Newman, Fairfield, Bhattacharya, & Grupenhoff, 1997).

There has been a swelling of research investigating the health benefits of a variety of bioactive compounds found in tree and plant extracts (Kris-Etherton & Hecker, 2002). Polyphenols, the most abundant antioxidants in the diet, have preventive properties for degenerative diseases such as cardiovascular disease and cancer (Scalbert, Johnson, & Saltmarsh, 2005).[11] Phytoestrogens, found most commonly in soy, are another class of bioactive compounds that have been associated with lowered risk of osteoporosis, heart disease, breast cancer, and menopausal symptoms (Patisaul & Jefferson, 2010). Like most things, the benefits of these compounds are negated with excessive consumption. Much more research is needed, but without the conservation of the GI that supports biodiversity, many bioactive compounds and their potential health benefits could be lost. Green infrastructure is a living laboratory of medicine. Finding these compounds may not just be a matter of putting every leaf, stem, and root on earth under the microscope. In her engaging book, *Biomimicry*, Jane Benyus (1997) makes a compelling case for observing the natural environment for its lessons on how to live healthier. In the case of medicine, there could be lessons learned from observing the behavior of ill animals that seek out certain plants for their healing qualities. Conserving GI and biodiversity allows the opportunity to study such actions.

An impressive compilation of information on the role of nature in the evolution of medicine, the significance of biodiversity in biomedical research, and the valuable organisms currently under threat can be found in *Sustaining Life: How Human Health Depends on Biodiversity* (Chivian & Bernstein, 2008) and in "Forest Products with Health-Promoting and Medicinal Properties" (Gallis et al., 2011). Karjalainen et al. (2010) also include a review of some of the medicinal benefits of GI as part of their broader review of the role of forests in promoting human health.

The popular reawakening in the western world to the healing properties of plants is evident in the explosion in the use of herbal medicines and medicinal plants that some one billion people worldwide use for healing (World Bank, 2004). While pharmaceuticals are essential in the treatment of some diseases, the enormous investment in biomedicine over the past 50 years has not translated into a parallel return on investment in health and well-being, and commonly counterproductive side effects have led people to go directly to the source of many pharmaceuticals, that being the plants from which many pharmaceuticals are derived. The Secretariat of the Convention on Biological Diversity estimated global sales of herbal products at $60 billion in 2000 (WHO, 2003). Eighty percent of the population in developing countries rely largely on herbal medicines for their health care needs, and the WHO estimates that a similar percentage of the world population may soon rely on herbal remedies (United Nations, 2004, p. 1).

The role of biodiversity in medicine has received a great deal of attention among nature and health research and advocacy. So much so that leading authors in this area have cautioned against allowing this focus to come at the expense of equally, if not more, fundamental ecosystem services (Newman, Kilama, Bernstein, & Chivian, 2008). The billions spent on herbal medicines and synthetic pharmaceuticals ceases to be pertinent if the natural environment cannot support other ecosystem services essential for life and health; there is no pill for fixing the environment. The pharmaceutical potential of GI is only one among the many health-supporting ecosystem services of GI. In the ecological era of public health, we now know better than to rely solely on the failed promise of biomedical solutions.

Summary

The protection of the landscape and GI is essential to support the basic ecosystem services on which human health depends. These services include the water, air, and food that form the foundation of life. Green infrastructure regulates both water quantity and quality. The role GI plays in regulating water quantity stems foremost from its role in the hydrological cycle and its ability to facilitate the recharge of groundwater stores. In this way, GI aids in the cycling and capturing of water that humans use for drinking and other essential needs, but GI also regulates water quantity by controlling surface runoff volumes. Green infrastructure regulates water quality through its ability to filter pollutants that fall with the rain and then collect in surface runoff. Green infrastructure not only removes pollutants from water but also from the air. Plants remove significant amounts of gaseous and particulate pollutants directly from the air, but they cannot remove these pollutants at anywhere near the levels at which these pollutants are being emitted into the atmosphere. Green infrastructure can improve air quality and respiratory health, but it is a not a stand-alone solution to counteract the current rate of emissions. Green infrastructure is vital to the food we eat because *all* the calories humans consume, no matter at which point on the food chain one focuses, are derived from the primary production of plants. Without photosynthetic organisms, there would be no life on earth. Green infrastructure is also critical to nutrient cycling and the pollination essential to food production. Water, air, and food are essential to maintaining life, but GI also protects biodiversity and, by doing so, the reservoir of medicinal compounds (both herbal and synthesized) that can improve health when it is compromised.

Notes

1. In our discussion of climate change and heat, we will also see how impervious surfaces contribute to local microclimates.
2. The ability of GI to support non-motorized forms of locomotion is discussed in Chapter 6, Physical Activity.
3. I was unable to discern over what period of time, but it is assumed to be a day.
4. Plants can remove airborne pollutants through the various chemical reactions that occur on plant surfaces, often with the aid of precipitation. The process by which trees remove pollutants from the air without the aid of precipitation is called dry deposition.
5. The compensatory value of the loss of urban trees has been estimated at $2.4 trillion in the USA (Nowak, Crane, & Dwyer, 2002; Nowak, Hoehn, Crane, Stevens, & Walton, 2007). Compensatory value is the cost to replace them if they were lost. Parts of the total dollar value are the perceived benefits of trees and the desire to have them present in one's environment.
6. Trees also remove ammonia (Tyrväinen et al., 2005), but this is not addressed here.
7. NO_x represents the family of nitrogen oxides including NO, NO_2, and nitrous and nitric acids. NO_2 is the compound of greatest interest to health and as an indicator of other NO_x compounds.
8. The VOC designation excludes CO, CO_2, and a handful of other organic compounds. VOCs are present in many everyday products and materials, and they are volatile because they are emitted from these products under normal

atmospheric conditions. VOCs are also present out-of-doors when released from manufacturing processes, but are most pertinent to our discussion of GI and air quality due to their role in ozone formation.

9. Not defined here as GI, indoor plants and their soil may also play a role in removing VOCs, such as benzene and toluene, commonly emitted from the solvents that coat indoor furnishings and building materials and used in cleaning agents (Liu, Mu, Zhu, Ding, & Crystal Arens, 2007; Orwell, Wood, & Burchett, 2006; Yang, Pennisi, Son, & Kays, 2009).

10. Other more ecologically sensitive methods of protecting crops from pests include introducing plants to diversify crops and repel pests. Plants have been used to visually camouflage crops, to dilute their attractive stimuli, and to repel pests chemically. Plants are being used to protect plants.

11. Polyphenols have been found in their most concentrated form in the knots of trees (Holmbom et al., 2007).

References

Aldridge, J., & Sempik, J. (2002). *Social and therapeutic horticulture: Evidence and messages from research.* Loughborough, UK: Centre for Child and Family Research, Loughborough University.

Appel, L. J., Sacks, F. M., Carey, V. J., Obarzanek, E., Swain, J. F., Miller, E. R., et al. (2005). Effects of protein, monounsaturated fat, and carbohydrate intake on blood pressure and serum lipids: Results of the OmniHeart randomized trial. *Journal of the American Medical Association, 294*(19), 2455–64.

Armstrong, D. (2000). A survey of community gardens in upstate New York: Implications for health promotion and community development. *Health & Place, 6*(4), 319–27.

Arnold, C., & Gibbons, J. (1996). Impervious surface coverage: The emergence of a key environmental indicator. *Journal of the American Planning Association, 62*(2), 243–58.

Barnicle, T., & Midden, K. (2003). The effects of a horticulture activity program on the psychological well-being of older people in a long-term care facility. *HortTechnology, 13*(March), 81–5.

Bealey, W. J., McDonald, A. G., Nemitz, E., Donovan, R., Dragosits, U., Duffy, T. R., & Fowler, D. (2007). Estimating the reduction of urban PM10 concentrations by trees within an environmental information system for planners. *Journal of Environmental Management, 85*(1), 44–58.

Beckett, K. P., Freer-Smith, P. H., & Taylor, G. (1998). Urban woodlands: Their role in reducing the effects of particulate pollution. *Environmental Pollution, 99*(3), 347–60.

Beckett, K. P., Freer-Smith, P. H., & Taylor, G. (2000). The capture of particulate pollution by trees at five contrasting urban sites. *Arboricultural Journal, 24*(2–3), 209–30.

Benyus, J. M. (1997). *Biomimicry: Innovation inspired by nature*. New York, NY: William Morrow.

Carlton, A., & Pinder, R. (2010). To what extent can biogenic SOA be controlled? *Environmental Science & Technology, 44*, 3376–80.

Cavanagh, J.-A. E., Zawar-Reza, P., & Wilson, J. G. (2009). Spatial attenuation of ambient particulate matter air pollution within an urbanised native forest patch. *Urban Forestry & Urban Greening, 8*(1), 21–30.

Chivian, E., & Bernstein, A. (2008). *Sustaining life: How human health depends on biodiversity*. New York, NY: Oxford University Press.

Defries, A. (1927). *The interpreter Geddes: The man and his gospel*. London: George Routledge & Sons.

Domm, J., Drew, R., Greene, A., Ripley, E., Smardon, R., & Tordesillas, J. (2008). Recommended urban forest mixtures to optimize selected environmental benefits. *EnviroNews, 14*(1), 7–10.

Donovan, G. H., Butry, D. T., Michael, Y. L., Prestemon, J. P., Liebhold, A. M., Gatziolis, D., & Mao, M. Y. (2013). The relationship between trees and human health: Evidence from the spread of the emerald ash borer. *American Journal of Preventive Medicine, 44*(2), 139–45.

Du, S., Kang, D., Lei, X., & Chen, L. (2007). Numerical study on adjusting and controlling effect of forest cover on PM10 and O3. *Atmospheric Environment, 41*(4), 797–808.

Dwight, R. H., Baker, D. B., Semenza, J. C., & Olson, B. H. (2004). Health effects associated with recreational coastal water use: Urban versus rural California. *American Journal of Public Health, 94*(4), 565–7.

Eilers, E. J., Kremen, C., Smith Greenleaf, S., Garber, A. K., & Klein, A.-M. (2011). Contribution of pollinator-mediated crops to nutrients in the human food supply. *PloS ONE, 6*(6), e21363.

Environment Canada. (2012). Something in the air. March 26. Retrieved April 5, 2014, from http://ec.gc.ca/biotrousses-biokits/default.asp?lang=En& n=7C8DC97C-9AFF-4E0D-B2B1

Escobedo, F. J., & Nowak, D. J. (2009). Spatial heterogeneity and air pollution removal by an urban forest. *Landscape and Urban Planning, 90*(3), 102–10.

Gallis, C., Di Stefano, M., Moutsatsou, P., Sarjala, T., Virtanen, V., Holmbom, B., et al. (2011). Forest products with health-promoting and medicinal properties. In K. Nilsson, M. Sangster, C. Gallis, T. Hartig, S. de Vries, K. Seeland, & J. Schipperijn (Eds.), *Forests, trees and human health* (pp. 41–76). New York, NY: Springer.

Golden, C. D., Fernald, L. C. H., Brashares, J. S., Rasolofoniaina, B. J. R., & Kremen, C. (2011). Benefits of wildlife consumption to child nutrition in a biodiversity hotspot. *Proceedings of the National Academy of Sciences, 108*(49), 19653–6.

Grifo, F., Newman, D., Fairfield, A., Bhattacharya, B., & Grupenhoff, J. (1997). The origin of prescription drugs. In F. Grifo & J. Rosenthal (Eds.), *Biodiversity of human health* (pp. 131–63). Washington, DC: Island Press.

Hartig, T., Mitchell, R., de Vries, S., & Frumkin, H. (2014). Nature and health. *Annual Review of Public Health, 35*, 21.1–21.22.

Hedin, L. (2000). Deposition of nutrients and pollutants to ecosystems. In O. E. Sala, R. B. Jackson, H. A. Mooney, & R. W. Howarth (Eds.), *Methods in ecosystem science* (pp. 265–76). New York, NY: Springer-Verlag.

Hillel, D., & Rosenzweig, C. (2008). Biodiversity and food production. In E. Chivian & A. Bernstein (Eds.), *Sustaining life: How human health depends on biodiversity* (pp. 325–81). New York, NY: Oxford University Press.

Holmbom, B., Willfoer, S., Hemming, J., Pietarinen, S., Nisula, L., Eklund, P., & Sjoeholm, R. (2007). Knots in trees: A rich source of bioactive polyphenols. In D. S. Argyropoulos (Ed.), *Materials, chemicals, and energy from forest biomass* (Vol. 954, pp. 350–62). Washington, DC: American Chemical Society.

Islam, M. N., Rahman, K.-S., Bahar, M. M., Habib, M. A., Ando, K., & Hattori, N. (2012). Pollution attenuation by roadside greenbelt in and around urban areas. *Urban Forestry & Urban Greening, 11*(4), 460–4.

Jack, D. W., & Kinney, P. L. (2010). Health co-benefits of climate mitigation in urban areas. *Current Opinion in Environmental Sustainability, 2*(3), 172–7.

Jim, C. Y., & Chen, W. Y. (2008). Assessing the ecosystem service of air pollutant removal by urban trees in Guangzhou (China). *Journal of Environmental Management, 88*(4), 665–76.

Karjalainen, E., Sarjala, T., & Raitio, H. (2010). Promoting human health through forests: Overview and major challenges. *Environmental Health and Preventive Medicine*, *15*(1), 1–8.

Karl, T., Harley, P., Emmons, L., & Thornton, B. (2010). Efficient atmospheric cleansing of oxidized organic trace gases by vegetation. *Science*, *330*(6005), 816–19.

Kearns, C., Inouye, D., & Waser, N. (1998). Endangered mutualisms: The conservation of plant–pollinator interactions. *Annual Review of Ecology and Systematics*, *29*(1998), 83–112.

Klein, A.-M., Vaissière, B. E., Cane, J. H., Steffan-Dewenter, I., Cunningham, S. A., Kremen, C., & Tscharntke, T. (2007). Importance of pollinators in changing landscapes for world crops. *Proceedings of the Royal Society Biological Sciences*, *274*(1608), 303–13.

Kling, G. (2008). The flow of energy: Primary production to higher trophic levels. Retrieved August 26, 2013, from www.globalchange.umich.edu/global-change1/current/lectures/kling/energyflow/energyflow.html

Krausmann, F., Erb, K.-H., Gingrich, S., Haberl, H., Bondeau, A., Gaube, V., et al. (2013). Global human appropriation of net primary production doubled in the 20th century. *Proceedings of the National Academy of Sciences*, *110*(25), 10324–9.

Kris-Etherton, P., & Hecker, K. (2002). Bioactive compounds in foods: Their role in the prevention of cardiovascular disease and cancer. *American Journal of Medicine*, *113*(Suppl 9B), 71S–88S.

Liu, Y.-J., Mu, Y.-J., Zhu, Y.-G., Ding, H., & Crystal Arens, N. (2007). Which ornamental plant species effectively remove benzene from indoor air? *Atmospheric Environment*, *41*(3), 650–4.

Lovasi, G. S., Quinn, J. W., Neckerman, K. M., Perzanowski, M. S., & Rundle, A. (2008). Children living in areas with more street trees have lower prevalence of asthma. *Journal of Epidemiology and Community Health*, *62*(7), 647–9.

Lovasi, G. S., O'Neil-Dunne, J. P. M., Lu, J. W. T., Sheehan, D., Perzanowski, M. S., Macfaden, S. W., et al. (2013). Urban tree canopy and asthma, wheeze, rhinitis, and allergic sensitization to tree pollen in a New York city birth cohort. *Environmental Health Perspectives*, *121*(4), 494–500.

Lukes, R., & Kloss, C. (2008). *Managing wet weather with green infrastructure. Municipal handbook, green streets.* Washington, DC: US Environmental Protection Agency.

McPherson, E., Nowak, D. J., & Rowntree, R. (1994). *Chicago's urban forest ecosystem:*

Results of the Chicago Urban Forest Climate Project. Radnor, PA: US Department of Agriculture, Forest Service, Northeastern Forest Experiment Station.

Melillo, J., & Sala, O. (2008). Ecosystem services. In E. Chivian & A. Bernstein (Eds.), *Sustaining life: How human health depends on biodiversity* (pp. 75–115). New York, NY: Oxford University Press.

Millennium Ecosystem Assessment. (2005). *Ecosystems and human well-being: Health synthesis*. Geneva: WHO.

Milligan, C., Gatrell, A., & Bingley, A. (2004). "Cultivating health": Therapeutic landscapes and older people in northern England. *Social Science & Medicine, 58*(9), 1781–93.

Minnesota Pollution Control Agency. (2000). *Protecting water quality in urban areas: Best management practices for dealing with storm water runoff from urban, suburban and developing areas of Minnesota*. Saint Paul, MN: Minnesota Pollution Control Agency.

Morse, R., & Calderone, N. (2000). The value of honey bees as pollinators of US crops in 2000. *Bee Culture*, (March), 1–15.

Myers, S. S., Gaffikin, L., Golden, C. D., Ostfeld, R. S., Redford, K. H., Ricketts, T. H., et al. (2013). Human health impacts of ecosystem alteration. *Proceedings of the National Academy of Sciences, 110*(47), 18753–60.

Naranjo, L. (2011). Volatile trees. Retrieved May 26, 2015, from https://earthdata.nasa.gov/featured-stories/featured-research/volatile-trees

National Library of Medicine. (2013). Nitrogen oxides. Retrieved May 26, 2015, from http://toxtown.nlm.nih.gov/text_version/chemicals.php?id=19

Newman, D., Kilama, J., Bernstein, A., & Chivian, E. (2008). Medicines from nature. In E. Chivian & A. Bernstein (Eds.), *Sustaining life: How human health depends on biodiversity* (pp. 117–61). New York, NY: Oxford University Press.

Nicholson-Lord, D. (2003). *Green cities—and why we need them*. London: New Economies Foundation.

Nowak, D. J. (1995). Trees pollute? A "TREE" explains it all. In M. Barratt & C. Kollin (Eds.), *Proceedings of the 7th National Urban Forestry Conference* (pp. 28–30). Washington, DC: American Forests.

Nowak, D. J., Civerolo, K. L., Trivikrama Rao, S., Luley, C. J., & Crane, D. E. (2000). A modeling study of the impact of urban trees on ozone. *Atmospheric Environment, 34*(10), 1601–13.

Nowak, D. J., Crane, D., & Dwyer, J. (2002). Compensatory value of urban trees in the United States. *Journal of Arboriculture, 28*(4), 194–9.

Nowak, D. J., Crane, D. E., & Stevens, J. C. (2006). Air pollution removal by urban trees and shrubs in the United States. *Urban Forestry & Urban Greening*, *4*(3–4), 115–23.

Nowak, D. J., Hoehn, R., & Crane, D. (2007). Oxygen production by urban trees in the United States. *Arboriculture and Urban Forestry*, *33*(3), 220–6.

Nowak, D. J., Hoehn, R., Crane, D., Stevens, J., & Walton, J. (2007). *Assessing urban forest effects and values New York City's urban forest*. Newton Square, PA: USDA Forest Service.

Nowak, D. J., Hirabayashi, S., Bodine, A., & Hoehn, R. (2013). Modeled PM2.5 removal by trees in ten U.S. cities and associated health effects. *Environmental Pollution*, *178*, 395–402.

Nowak, D. J., Hirabayashi, S., Bodine, A., & Greenfield, E. (2014). Tree and forest effects on air quality and human health in the United States. *Environmental Pollution*, *193*, 119–29.

Oliver-Solà, J., & Núñez, M. (2007). Service sector metabolism: Accounting for energy impacts of the Montjuïc urban park in Barcelona. *Journal of Industrial Ecology*, *11*(2), 83–98.

Orwell, R., Wood, R., & Burchett, M. (2006). The potted-plant microcosm substantially reduces indoor air VOC pollution: II. Laboratory study. *Water, Air, and Soil Pollution*, *177*, 59–80.

Pataki, D. E., Carreiro, M. M., Cherrier, J., Grulke, N. E., Jennings, V., Pincetl, S., et al. (2011). Coupling biogeochemical cycles in urban environments: Ecosystem services, green solutions, and misconceptions. *Frontiers in Ecology and the Environment*, *9*(1), 27–36.

Patisaul, H., & Jefferson, W. (2010). The pros and cons of phytoestrogens. *Frontiers in Neuroendocrinology*, *31*(4), 400–19.

Perraud, V., Bruns, E. A., Ezell, M. J., Johnson, S. N., Yu, Y., Alexander, M. L., et al. (2012). Nonequilibrium atmospheric secondary organic aerosol formation and growth. *Proceedings of the National Academy of Sciences*, *109*(8), 2836–41.

Pimentel, D., Wilson, C., McCullum, C., Huang, R., Dwen, P., Flack, J., et al. (1997). Economic and environmental benefits of biodiversity. *BioScience*, *47*(11), 747–57.

Pope, C. A., Burnett, R. T., Thun, M. J., Calle, E. E., Krewski, D., & Ito, K. (2002). Lung cancer, cardiopulmonary mortality and long-term exposure to fine particulate air pollution. *Journal of the American Medical Association*, *287*(9), 1132–41.

Sampat, P. (2000). Groundwater shock. *World Watch*, (January/February), 10–22.

Sanesi, G., Gallis, C., & Kasperidus, H. D. (2011). Urban forests and their ecosystems services in relation to human health. In K. Nilsson, M. Sangster, C. Gallis, T. Hartig, S. de Vries, K. Seeland, & J. Schipperijn (Eds.), *Forests, trees and human health* (pp. 23–40). New York, NY: Springer-Verlag.

Scalbert, A., Johnson, I. T., & Saltmarsh, M. (2005). Polyphenols: Antioxidants and beyond. *American Journal of Clinical Nutrition*, *81*(1 Suppl.), 215S–217S.

Shafer, C. S., Lee, B. K., & Turner, S. (2000). A tale of three greenway trails: User perceptions related to quality of life. *Landscape and Urban Planning*, *49*(3–4), 163–78.

Shinew, K. J., Glover, T. D., & Parry, D. (2004). Leisure spaces as potential sites for interracial interaction: Community gardens in urban areas. *Journal of Leisure Research*, *36*(3), 336–55.

Taub, D. R., Miller, B., & Allen, H. (2008). Effects of elevated CO_2 on the protein concentration of food crops: A meta-analysis. *Global Change Biology*, *14*(1608), 565–75.

TED. (2013). A guerilla gardener in South Central LA. Retrieved January 30, 2015, from www.ted.com/talks/ron_finley_a_guerilla_gardener_in_south_central_la?language=en

Tranel, M., & Handlin, L. B. (2006). Metromorphosis: Documenting change. *Journal of Urban Affairs*, *28*(2), 151–67.

Tyrväinen, L., Pauleit, S., Seeland, K., & de Vries, S. (2005). Benefits and uses of urban forests and trees. In C. Konijnendijk, K. Nilsson, T. Randrup, & J. Schipperijn (Eds.), *Urban forests and trees: A reference book* (pp. 81–114). Berlin: Springer.

United Nations. (2004). *Trade in medicinal plants*. Rome: Food and Agriculture Organization of the United Nations.

US EPA. (1995). Particulate matter (PM-10). Retrieved May 26, 2015, from www.epa.gov/airtrends/aqtrnd95/pm10.html

US EPA. (2000). *National air pollutant emission trends: 1900–1998 report*. Washington, DC: US Environmental Protection Agency.

US EPA. (2013a). Air emission sources. Retrieved October 24, 2013, from www.epa.gov/cgi-bin/broker?_service=data&_debug=0&_program=dataprog.national_1.sas&polchoice=SO2

US EPA. (2013b). Health effects of ozone in the general population. Retrieved September 11, 2013, from www.epa.gov/apti/ozonehealth/population.html

US EPA. (2015). Sulfur dioxide. Health. Retrieved May 26, 2015, from www.epa. gov/oaqps001/sulfurdioxide/health.html

US Geological Service. (2013). The water cycle: Evapotranspiration. Retrieved April 11, 2013, from http://ga.water.usgs.gov/edu/watercycleevapotranspiration.html

van Essen, H., Schroten, A., Otten, M., Sutter, D., Schreyer, C., Zandonella, R., et al. (2011). *External costs of transport in Europe*. Delft, the Netherlands: CE Delft, Infras, Fraunhofer ISI.

Vitousek, P., Ehrlich, P., Ehrlich, A., & Matson, P. (1986). Human appropriation of the products of photosynthesis. *BioScience, 36*(6), 368–73.

Whitlow, T. H., Pataki, D. A., Alberti, M., Pincetl, S., Setala, H., Cadenasso, M., et al. (2014). Comments on "Modeled PM2.5 removal by trees in ten U.S. cities and associated health effects" by Nowak et al. (2013). *Environmental Pollution, 191*, 256.

WHO. (2003). *WHO guidelines on good agricultural and collection practices for medicinal plants*. Geneva: World Health Organization.

Wolf, K. (2004). *Trees, parking and green law: Strategies for sustainability*. Stone Mountain, GA: Georgia Forestry Commission, Urban and Community Forestry.

World Bank. (2004). *Sustaining forests: A development strategy* (Vol. 1). Washington, DC: World Bank.

Yang, D., Pennisi, S., Son, K., & Kays, S. (2009). Screening indoor plants for volatile organic pollutant removal efficiency. *HortScience, 44*(5), 1377–81.

Yang, J., McBride, J., Zhou, J., & Sun, Z. (2005). The urban forest in Beijing and its role in air pollution reduction. *Urban Forestry & Urban Greening, 3*(2), 65–78.

Zanobetti, A., & Schwartz, J. (2009). The effect of fine and coarse particulate air pollution on mortality: A national analysis. *Environmental Health Perspectives, 117*(6), 898–903.

Chapter Four

The challenge of climate change

Although there still remains some dissent over the extent to which humans are responsible for climate change, this should in no way detract from the absolute certainty that climatic norms are changing. Expert opinion largely concurs that, with a changing climate, the level and type of disease burden in all countries and regions will progressively increase (Confalonieri et al., 2007). Our nascent understanding of how global health already has been, and will continue to be, affected by climate change reveals that GI, yet again, plays a critical part in improving the human condition (Amati & Taylor, 2010). As climatic conditions change, GI will change, and so too will the ecosystem services on which health depends. Alterations to the landscape not sensitive to GI exacerbate the flux in climatic conditions, and any assessment of how climate change will impact health must include the alterations that humans make to the landscape (Patz, Campbell-Lendrum, Holloway, & Foley, 2005, p. 315). Green infrastructure can both mitigate climate change and support human adaptation to changing climatic conditions and events.

Since this chapter is intentionally limited to the health outcomes of climate change that GI can alleviate, it overlooks a number of other potential health effects. I therefore recommend three sources of further reading on the topic. First, a Lancet commission report on *Managing the Health Effects of Climate Change* outlines the health threats of climate change and also the policy and institutional responses, including improved land-use regulations, to address these threats (Costello et al., 2009). Second, a similar report, *A Human Health Perspective on Climate Change*, provides the most extensive litany to date of the potential health effects of climate change (National Institute of Environmental Health Sciences, 2010). Third, *Globalization, Climate Change, and Human Health*, diagrams the complex pathways by which climatic changes are linked to health outcomes and provides recommended health sector responses (McMichael, 2013).

There has been a small surge in the literature in recent years exploring the potential health impacts of global climate variability. Although the body of literature is growing, guidance on this subject is still sparse, and the health outcomes that have been studied have largely been limited to those associated with heatwaves and air pollution (Ebi, 2010). While this may be a product of the routine use of standard indicators of environmental health, there has been a recognition that a number of other environmental indicators (e.g. use of renewable energy) are needed alongside more traditional indicators to measure the health effects of climate change (English et al., 2009). Table 4.1 is a synopsis of the litany of potential health effects of climate change.

Of course, projecting the level of increased morbidity and mortality caused by climate change is extremely complex and fraught with uncertainty. Nonetheless, there have been some efforts by public health researchers to quantify the added disease burden attributable to the conditions that may result from a changing climate. With a recognized appreciation for the ecological complexity of making such predictions, it has been shown that climate change has already affected human health and that the associated risks from a number of conditions will likely increase over time (McMichael et al., 2003, Table 7.2, p. 140). In the year 2000 alone, climate variability was likely responsible for over 150,000 deaths worldwide with almost 90 percent of this increase in the disease burden falling upon children (McMichael et al., 2004). There has been some justified criticism of the methods used to achieve this estimate and the small proportion this added mortality is to global deaths (0.3 percent) (Goklany, 2007). Despite this, a changing climate is undoubtedly having a negative effect on health, and it is believed that time will only make this proportion grow with certain populations being hit much harder than others. The particular burden placed on childhood populations by anticipated

Table 4.1: Summary of health effects of climate change

Climatic event	Intermediary	Health outcome
Heat waves	direct to ➔ Increased ground-level ozone, pollen	Heat stress, stroke Respiratory disease exacerbation
Increased mean temperature	direct to ➔ More hospitable to disease vectors (e.g. mosquito, ticks) More hospitable to infectious disease agents (e.g. bacteria)	Positive: Less hypothermia Vector-borne diseases (e.g. Lyme, malaria, dengue) Food-poisoning, infectious disease (e.g. cholera)
Ozone depletion	UV radiation	Skin and eye maladies
Drought	Water/food shortage Lack of water safety	Dehydration, malnutrition Water-borne disease
Extreme weather event (e.g. flooding, tornado, hurricane)	direct to ➔ Population movement Lack of food/water safety	Injuries, drowning Conflicts Water-borne disease, malnutrition
Sea-level rise	direct to ➔ Population movement Water/soil salinization	Injuries, drowning Conflicts Dehydration, malnutrition
Climate change generally	Stress	Mental health

Source: Adapted from Coutts and Berke (2013).

Note: Compiled with data from Confalonieri et al. (2007); Frumkin, Hess, Luber, Malilay, and McGeehin (2008); McMichael, Woodruff, and Hales (2006); and Costello et al. (2009).

climatic events has been the focus of a handful of studies (e.g. Kistin, Fogarty, Pokrasso, McCally, & McCornick, 2010; Sheffield & Landrigan, 2011).

Disability–Adjusted Life Year (DALY) has also been applied to the health effects of climate change to take "into account not only the proportional change in each impact, but also the size of the disease burden" (McMichael et al., 2003, p. 154). Disability–Adjusted Life Year combines the duration and severity of disability and years of life lost due to premature mortality to reveal the disease burden in a population. A conservative estimate—taking into account cardiovascular disease deaths, diarrhea episodes, malaria and dengue cases, fatal unintentional injuries, and malnutrition—is that 5.5 million DALYs were lost globally in 2000 due to the effects of climate change (Campbell-Lendrum, Corvalán, & Prüss-Ustün, 2003).[1]

There have also been efforts to capture the economic costs of the health effects of climate change. In Europe, the most conservative estimates reveal that, although food-borne and mental illnesses are expected to rise, the most significant costs between 2011 and 2040 will be the billions of euros spent to address rises in heat- and cold-related mortality (Watkiss, Horrocks, Pye, Searl, & Hunt, 2009). Again, there are those who have reasonably cautioned against an alarmist attitude regarding the health effects of climate change (Rohr et al., 2011). Nonetheless, the nascent literature connecting climate change to health outcomes offers some guidance on the impending threats to the populations public health is striving to protect, and a precautionary approach should lead the public health sector to take the threats to health and associated costs seriously.

McMichael (2012), while also offering an insight into the historical lessons derived from the health effects of a constantly changing climate, organizes the health impacts of climate change into three categories.

1. Direct impacts: heatwaves, floods, extreme weather events, air pollutants
2. Changes in ecological or biophysical systems: food yields, potable water, infectious disease
3. Social and economic disruptions: diverse health disorders due to forced migration, depression and despair in marginalized groups, hostilities due to climate-related declines in ecosystem services.

In the upcoming sections we examine how GI can assist in ameliorating: (1) direct impacts, and (2) changes to ecological and biophysical systems (heat, weather events, infectious disease). In the preceding chapter we have explored how GI is vital to the ecological and biophysical systems that provide water, air, and food. As the climate changes, so too will GI and the ecological systems that provide these life-supporting elements. The third category, social and economic disruptions, are not addressed here in any depth, but if the direct impacts and changes to ecological systems are reduced by conserving GI, then it is likely that there will be fewer social and economic disruptions caused by persons competing for increasingly scant resources. Green infrastructure plays a role in mitigating impacts in all three of these categories due to its ability to slow climate change, a measure of primary prevention. It also has a role to play in adapting to impending climate change by diminishing the severity of events that pose risks to health.

Green infrastructure and carbon sequestration: mitigation and primary prevention

Green infrastructure, and in particular trees, can mitigate the potential adverse health effects caused by climate change due to the ability of plants to capture and store carbon from the greenhouse gas CO_2. In 2011, CO_2 accounted for 84 percent of the total greenhouse gas emissions in the USA (US EPA, 2013c), the greatest source of these emissions coming from the combustion of petroleum, coal, and natural gas. Green infrastructure can aid in cooling the local and global environment by sequestering carbon during the photosynthesis process.

Even if all CO_2 emissions were to stop today, we would be dealing with the ramifications of current concentrations of carbon in the atmosphere for at least the next 40 years and possibly much longer. The climate change that will occur over the next 30 to 40 years has already been set in motion by the carbon dioxide currently in the atmosphere (Hulme et al., 2002). While it is absolutely necessary to reduce emissions to prepare for life beyond 40+ years, it is also advantageous to health to capture or sequester the carbon already in the atmosphere. A study of 10 major US cities found that the gross carbon sequestration rate of urban trees was 22.8 metric tons of carbon/year (a $460 million/year value) (Nowak & Crane, 2002). Nationally in the USA, net carbon sequestration in the forest sector in 2005 offset 10 percent of CO_2 emissions (Woodbury, Smith, & Heath, 2007).

Larger and more mature trees are able to store more carbon than younger trees, but this does not necessarily mean that pristine, old-growth forests are the most efficient at capturing carbon. In central California, observed rates of annual carbon sequestration ranged from 35 pounds for small trees to 800 pounds for large, mature trees (Wolf, 2004), but younger trees in rapid growth stages capture carbon at a higher rate than their more mature and slower-growing counterparts. Also, in mature forests, carbon sequestration is counterbalanced by the decomposition and release of carbon from trees that have reached the end of their lifespan. The selective harvesting of trees for wood products keeps carbon stored in the dry wood (where it makes up half of the material mass) for a longer period as compared to if trees were allowed to decay, but the tree residues and roots left behind to decay after harvesting may offset gains in the dry wood storage of carbon. Harvested wood burned for fuel releases sequestered carbon back into the atmosphere immediately, but burning wood for fuel is a much more carbon neutral option as compared to fossil fuel combustion. Indeed, it is the carbon released from fossil fuel combustion that we are counting on trees to capture. The ability of forests to act as net carbon sinks is complicated by changing climatic conditions, soil fertility, varied carbon sequestration rates of tree species, and a range of other issues that are not completely understood (Bellassen & Luyssaert, 2014). The lesson here is that when considering the potential of trees and forests to act as carbon sinks, the sustainable harvesting of forest products for fuel and building materials may not be contrary to efficient carbon sequestration, climate change mitigation, and health promotion.

Despite the impressive ability of trees to capture and store carbon, planting trees is not the panacea for mitigating the climatic change caused by increasing levels of atmospheric CO_2. Efforts to increase the amount and quality of GI for carbon capture will need to be accompanied by a reduction in emissions. According to the latest climate-change science used by the 2009 US Global Change Research Program to project national and regional climate-change effects, in the absence of preventive and mitigating measures aimed at reducing CO_2 emissions, the present atmospheric CO_2 concentration of about 385 parts per million (ppm) could grow by 50 percent (low emission scenario) to more than 950 ppm (highest emissions scenario) by the end of the century (Karl, Melillo, & Peterson, 2009). The public appears to be aware of the need for this type of multifaceted solution aimed at not only capture but also emission reduction. A study in Canada revealed that a greenhouse gas reduction strategy is likely to be accepted as long as it is used in combination with energy efficiency and alternative energy technologies (Sharp, Jaccard, & Keith, 2009). Actions such as expanding GI will only be effective, and are only likely to be accepted, if they are done while simultaneously pursuing alternative energy technologies that reduce emissions.

The important role GI plays in capturing carbon and mitigating climate change is evident in its ability to reduce the emissions produced from heating and cooling buildings. In the next section on GI as an adaptation strategy, I will discuss how GI can help reduce the health effects of rising ambient heat (particularly in cities), but GI is also a mitigation strategy since it can capture carbon and reduce the carbon emitted when artificially heating and cooling structures. Trees reduce the amount of emissions produced to heat buildings by reducing wind speed. In the summer months, the same trees shade and cool buildings. Nowak (1993) estimated that planting 100 million urban trees could capture 77 million metric tons of carbon (tC) and avoid the production of 286 million tC from power plants over 50 years. Unfortunately, although significant in absolute tonnage (363 million tC), this was less than 1 percent of the estimated amount of carbon emitted over the same 50-year period. This would be equivalent to increasing the fuel efficiency of passenger automobiles by 0.5 km/l.

The relationship between GI and reducing the carbon emissions from buildings must also include larger considerations to the distribution of structures across the landscape. There must be a balance between introducing the GI that sequesters carbon and reduces the release of carbon and increasing the density of urban development. Development that spreads structures (and people living in those structures) across the landscape in low densities (e.g. the suburbs) has a larger carbon footprint than more dense development (Jones & Kammen, 2014).[2] In other words, individuals spread across the landscape in lower density yet greener environments produce more carbon than persons in denser, grayer environments. Therefore, the solution to mitigate the release of carbon may not to be to surround individual homes with acres of GI but rather to add the myriad benefits of GI to denser built environments that have reduced carbon footprints. Greening the denser gray built environment also protects GI outside the city. Green infrastructure outside the city can aid in the sequestration of the reduced per capita carbon releases from inside the city. The final piece of this puzzle would then be to connect a city's GI to regional GI. Carbon sequestration and other ecosystem services would be supported by the presence of GI outside of cities, but other health benefits would also be maximized from the increased access and exposure to a locally and regionally connected GI system.

Weighing the benefits of urban GI can be measured in monetary terms. This has been done by comparing the values of the carbon sequestration, energy reduction, and air pollution mitigation services of GI. Canberra, Australia, is a case in point. Over the course of a century, nearly half a million trees were planted in Canberra. The combined value of the energy reduction, pollution mitigation, and carbon sequestration services over a four-year period (2008–12) was US $20–67 million, but pulling these services apart it was found that "the dollar value of carbon sequestered in the urban forest [was] low relative to the estimated value of avoiding energy consumption and ameliorating air-borne and water pollution" (Brack,

2002, p. S199). Carbon sequestration is a valuable service, but not as valuable as the money saved from reducing the energy use that produced the carbon and that must then be sequestered. In this study it is not clear if the dollar value placed on ameliorating airborne and water pollution included the health care costs avoided due to a reduced exposure to these pollutants. If it does not, the true value of these urban trees is likely underrepresented. Health costs could also be tied to energy consumption by estimating the reduction of exposure to pollutants not released due to trees regulating ambient temperatures. The local impacts on health stemming from the ability of local GI to sequester carbon are likely to be minimal (especially considering rates of emissions), but the contribution of local GI in the sum of our global GI inventory is likely to have a great effect on the global phenomenon of climate change and the global health consequences of these changes.

Lastly, by not including GI in climate change mitigation measures aimed at capturing carbon and reducing its release, it is GI itself that is at risk. Climate change, deforestation, and desertification feed off one another in a slowly evolving loop. In addition to the immediate threats deforestation has on the loss of products with economic and salutogenic value, deforestation also erases the carbon capture and storage capacity of forests. Forests not only capture and store carbon produced by human activities, but they also capture and store the carbon released into the atmosphere from decomposing plants. Deforested areas fail to capture the carbon from these two sources, and this has a cumulative effect on warming that kills the trees that remain. More carbon is released from the decomposing plants killed by warming, less carbon is captured and stored, and so on until desertification results. This is exacerbated by the fact that deforestation can reduce the regional rainfall needed to support trees and other GI (Nobre, Sellers, & Shukla, 1991).

Climate change will continue to alter the GI, terrestrial ecosystems, and ecosystem services on which humans depend.[3] The climate is changing, and mitigation measures that include GI should be viewed as one among many strategies aimed at slowing, but not necessarily halting, the process. Since the climate will continue to change despite our most aggressive mitigation efforts, it is also essential to devise methods to adapt, and it is GI again that can be used as an adaptation strategy to protect humans from impending climatic events.

Green infrastructure and weather and climatic events: adaptation and secondary prevention

The ability of GI to sequester carbon can reduce the pace and severity of climatic changes and their associated deleterious health outcomes. This will aid in mitigating the threats from future changes, but much of these changes are already underway, and adaptation is therefore a necessary complement to mitigation. With a changing global climate will come changes in localized weather, and current predictions warn of an increased frequency and severity of potentially harmful weather events. In 2012 there were 905 major natural disasters worldwide with 93 percent of these disasters being extreme weather-related (storms, floods) and climatological events (heatwaves, cold waves, droughts, and wildfires). Together these events accounted for 93 percent of the estimated 9,600 reported fatalities from natural disasters worldwide (Munich RE, 2013). In line with climate variability predictions, the number of major weather-related disasters has been climbing from an average of less than two per year in 1950 to more than six per year in 2007 (Munich RE, 2007). It is not just more people being put at risk due to growing populations and migration to risk-prone areas, but rather an increase in the frequency of these events. Conserving GI is an adaptation strategy that can protect people from the health consequences stemming from these events. This is particularly important in coastal zones.

The combination of the global migration to cities and the increase in the frequency of extreme weather events is cause for concern. Half of the world's major cities are located in coastal zones. These areas are nearly three times as dense as inland areas, and they are growing rapidly (Millennium Ecosystem Assessment, 2005). As coastal cities grow, what is often lost is the GI that can act as a natural buffer to coastal weather events. The destruction of coastal mangroves, forests and wetlands, coral reefs, and vegetated dunes has serious consequences for the 40 percent of humanity who reside within 100km of ocean shorelines and at less than 50m above sea level. The natural features being lost can increase the severity of the flooding, storm surge, and landslide activity that occurs during extreme storms (Cockburn, St. Clair, & Silverstein, 1999). Noteworthy is the fact the protection provided by the conservation of coastal vegetation (such as mangroves, Figure 4.1) against potentially destructive wave action is not linear but rather increases exponentially the more area that is conserved (Barbier et al., 2008).

Figure 4.1: Mangrove coastal barrier, Cayenne, French Guiana
Source: Antoine Hubert.

This relationship does level off. This is good news in that the maximum protective benefits of GI conservation can be identified and then balanced at a point at which development can occur without losing protective benefits.

Of course, it is not just the direct health effects and conditions caused by flooding and storm surge that are cause for concern. The social and economic disruptions, the third category described by McMichael (2012) (noted previously), caused by extreme weather events also bring with it a host of other cascading health issues associated with food production, infectious disease transmission, forced migrations, and the stress associated with such events. The health outcomes of these events, some surfacing years later, are not captured in estimates of human health impacts.

Higher average temperatures and heatwave events also lead to social disruptions and a number of cascading effects. In fact, heat and heatwaves are a much greater risk to health than other types of catastrophe. Green infrastructure has the ability to mitigate and slow future increases in local and global temperatures through its ability to sequester CO_2, but conserving GI is also an adaptive strategy that can immediately reduce ambient temperatures and the associated impacts on health.

There is no question that the globe is warming. The first 13 years of the twenty-first century have accounted for 13 of the 14 warmest years of global average surface temperatures on record (since 1850). Each of the last three decades have been warmer than the previous decade, culminating in 2001–10 as the warmest decade to date (World Meteorological Organization, 2014). As ambient temperatures rise, so too will the risk from the direct effects of heat on physiological processes. Some direct effects are less severe—heat cramps, heat edema (swelling), heat syncope (fainting)—while others such as heat exhaustion and heat stroke can cause organ damage and death (Allen & Segal-Gidan, 2007; US EPA, 2003). Between 1979 and 2003 in the USA, the 8,015 deaths attributed to heat does not make it stand out among other public health issues, but it is still more severe than other natural events that often receive much more media attention. More people died from extreme heat over this 24-year period than from hurricanes, lightning, tornadoes, floods, and earthquakes combined (CDC, 2009). While this is likely startling to some, it may be made even more startling by the fact that many more persons die annually from the secondary effects of heat and illnesses exacerbated by heat including cardiovascular and kidney disease. A conservative estimate is that almost five times as many persons die annually in large US cities taking into account these secondary effects, and it is expected that these numbers will rise even when accounting for individual acclimatization (Kalkstein & Greene, 1997). In the UK, the 2,000 annual deaths attributed to heat is expected to rise by around 257 percent by the 2050s (Hajat, Vardoulakis, Heaviside, & Eggen, 2014). While the annual number of deaths per year directly related to heat are minimal in public health terms, we are continually uncovering the ways that heat is responsible for illnesses made worse by heat. For example, heat and humidity have been found to increase the risk of stroke hospitalizations (Lichtman, Leifheit-Limson, & Goldstein, 2014).

It is also true that a gradually warming planet could be good for health if it reduces morbidity and mortality associated with cold weather. Overall, mortality is higher in cold weather months compared to warm weather months taking into account both the direct effects (e.g. hypothermia) and the indirect effects (e.g. increased rates of pneumonia, influenza, and other respiratory illnesses)

(McGeehin & Mirabelli, 2001). Estimates that have taken a broader categorization of heat/drought and winter weather have found that mortality associated with heat and cold is nearly the same with heat being only slightly more dangerous (Borden & Cutter, 2008). Of course, the measurable health outcomes of heat and cold are influenced by individual and collective adaptive behaviors. Taking adaptions to a warming environment into account, the predicted trend is that climate-related mortality in the summer months will increase and climate-related mortality in the winter will decrease slightly, resulting in a net increase in climate-related mortality in US cities (Kalkstein & Greene, 1997). Winter deaths will be reduced, but the increase in summer deaths will surpass these gains. In countries that do not normally experience extreme cold and heat, such as the UK, cold-related mortality is expected to decline by 2 percent by 2050 with this burden still remaining higher than heat-related mortality (not taking into account individual adaptive behaviors) (Hajat et al., 2014). It is certainly not clear if this prediction will hold if heatwave events continue to produce higher than expected mortality. This was evidenced in western Europe in 2003 when an unprecedented heatwave was estimated to have killed nearly 50,000 people (Kosatsky, 2005), and again in 2013 with thousands more deaths than were normally expected. Cold will likely still kill more people than heat in the coming decades (depending on how you measure it), but what is clear is that heat risk is on the rise while cold-related risk is receding.

Urban heat

Green infrastructure as an adaptive strategy is likely to have a pronounced impact on urban environments where most of the global population now lives and where temperatures are consistently higher. First recorded in 1833, the phenomenon of cities having higher temperatures than the surrounding countryside has now been demonstrated beyond a doubt (Gill, Handley, Ennos, & Pauleit, 2007; Oke, 1982). This *urban heat island* effect (Figure 4.2) remains the most intensively studied climatic feature of cities and a major focus among a much wider field of urban climatology (Arnfield, 2003; Souch & Grimmond, 2006). The US Environmental Protection Agency reports that "the annual mean air temperature of a city with 1 million people or more can be 1–3°C (1.8–5.4°F) warmer than the surrounding region" (US EPA, 2013a). Other estimates report a temperature difference as high as 7°C (12.6°F) (Wilby, 2003).

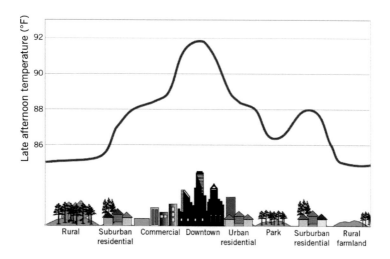

Figure 4.2: Transect of the urban heat island
Source: Adapted from Lemmen and Warren (2004).

The heat island effect is caused by the absorption of short-wave radiation by structures and ground-level artificial surfaces and increased concentrations of air pollution that absorb long-wave radiation. The roads, sidewalks, parking lots, and buildings concentrated in cities cumulatively create vast areas of absorptive surfaces. In the USA, only about one-fifth of all roads are urban, but when urban roads are taken together with other absorptive surfaces (e.g. sidewalks, parking lots), urban areas form intensively absorptive zones accounting for nearly 60 percent of the total absorptive surface area nationally (Lukes & Kloss, 2008). The low solar reflectance of roads and parking lots, typically composed of asphalt (5 to 10 percent solar reflectance), and their watertight characteristics that preclude solar heat dissipation through evaporation are the two factors that make roads and parking lots contributors to the urban heat island. In peak summer weather, unshaded asphalt areas can reach temperatures greater than 66°C (150.8°F). Given the heat storage capacity of such surfaces, unshaded roads and parking lots can raise surrounding air temperature by 11–22°C (19.8–39.6°F) (Wolf, 2004). This not only makes the environment feel hotter, but it also increases the demand for electricity to cool buildings. It is energy demand for air conditioning in buildings and in cars that generates more of the air pollutants that trap heat, further exacerbating the heat island.

It does not require large changes in temperature to translate into large increases in demand for energy. For every 0.6°C (1°F) increase in summertime temperature, peak utility loads in medium and large cities increase by an estimated 1.5–2.0 percent (US EPA, 2013b). So, if urban areas are approximately 3°C (5.4°F) hotter (a conservative estimate), this could translate into the heat island being responsible for 10 percent of the peak energy demand (Akbari, Pomerantz, & Taha, 2001; Akbari, 2005). This increase in temperature and energy demand can be partially ameliorated with GI. Green infrastructure in the form of shade trees can create a seasonal (~3 months in temperate zones) cooling energy savings of 30 percent (Akbari, Kurn, Bretz, & Hanford, 1997). Undoubtedly, a 30 percent reduction is substantial not only in monetary terms but also in the reduction in emissions which in turn exacerbate ambient heat further.

The lack of GI and water features in cities contributes to the urban heat island effect (Gill et al., 2007; Oke, 1982; Smith & Levermore, 2008). Green infrastructure can prevent the absorption of radiation by surfaces, the release of pollutants, and also cool the air through evapotranspiration. Parks within a city can have a significant cooling effect on local temperatures (Spronken-Smith & Oke, 1999). Even adding as little as 10 percent of greenspace in the form of trees in high-density development can reduce local surface temperatures by 1.4°C (2.5°F) on average (Pauleit & Duhme, 2000a, 2000b). Translating this into absolute terms and appreciating the fact that different forms of vegetation will vary slightly in their cooling potential, park size should exceed one hectare for significant cooling benefits, with at least 10 hectares needed to achieve a 1°C (1.8°F) reduction in air temperature (Kuttler, 1993 in German as cited by Tyrväinen, Pauleit, Seeland, & de Vries, 2005, p. 94). This reduction, although seemingly small, can nearly negate some urban heat islands. It would also keep the projected maximum surface temperatures caused by global warming up to the year 2080 at current levels (Gill et al., 2007) essentially negating the local temperature effects of global warming for the next 65 years. Shading absorptive surfaces with vegetation reduces the heat island, but shade also directly protects individuals from ultraviolet (UV) radiation. An individual tree can provide a Sun Protection Factor (SPF) of 6 to 10, a level of exposure to UV radiation one-sixth to one-tenth of full sun (NUFU, 1999 as cited by Tyrväinen et al., 2005, p. 86). This is important for not only preventing health issues associated with heat but also other associated health problems (Heisler, Grant, Grimmond, & Souch, 1995) such as skin cancer.

The evapotranspiration process discussed in Chapter 3 also plays an important role in regulating urban heat because moisture in the air reduces ambient temperatures. Now, you might be thinking, "Wait a minute, humidity makes me feel hotter." For human bodies this is true. As relative humidity increases, the ability of the body to cool itself through its own evaporative process (sweating) decreases. The ability of GI to decrease ambient temperatures by adding moisture to the air (evapotranspiration) is only effective up to a certain measure of relative humidity. This is why evaporative cooling devices are only effective in desert climates. As Figure 4.3 demonstrates, even at the low level of relative humidity of 40 percent, as the temperature rises above 90 degrees the heat index (how hot it feels) increases along with the risk of heat-related illness. This relationship becomes much more detrimental to health as the relative humidity increases. Therefore, the cooling effects of evapotranspiration are really only relevant to cooling individuals in climates with low relative humidity such as the Persian Gulf states, the interior of Australia, and the Southwestern USA. That said, it is not clear that it is a good idea to try to cool these types of environments with GI. Adding vegetation to these dry climates for the cooling properties of shade and evapotranspiration may bring with it a host of other ecologically disruptive issues. Evapotranspiration has a

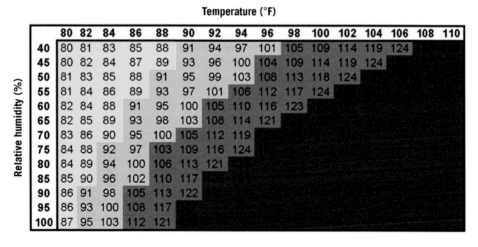

Temperature (°F)

	80	82	84	86	88	90	92	94	96	98	100	102	104	106	108	110
40	80	81	83	85	88	91	94	97	101	105	109	114	119	124		
45	80	82	84	87	89	93	96	100	104	109	114	119	124			
50	81	83	85	88	91	95	99	103	108	113	118	124				
55	81	84	86	89	93	97	101	106	112	117	124					
60	82	84	88	91	95	100	105	110	116	123						
65	82	85	89	93	98	103	108	114	121							
70	83	86	90	95	100	105	112	119								
75	84	88	92	97	103	109	116	124								
80	84	89	94	100	106	113	121									
85	85	90	96	102	110	117										
90	86	91	98	105	113	122										
95	86	93	100	108	117											
100	87	95	103	112	121											

Relative humidity (%) (vertical axis label, left side)

Likelihood of heat disorders with prolonged exposure or strenuous activity

☐ Caution   ☐ Extreme caution   ☐ Danger   ☐ Extreme danger

Figure 4.3: Heat index

Source: US National Oceanic and Atmospheric Association.

greater benefit to cooling buildings rather than cooling people, but there is still an ambient heat-reduction benefit to people. Cooler buildings consume less energy for artificial cooling and therefore produce less heat-trapping emissions.

Green roofs are an important form of urban GI that can reduce the heat island effect. Roofs are an often ignored form of urban real estate, but they have gained some attention in recent decades for their untapped potential to enhance environmental quality as it relates to storm water, gardening, and the urban heat island (Figure 4.4). Similar to ground-level GI, green roofs can ameliorate urban heat by negating the absorptive properties of rooftop surfaces and by reducing the energy demand to cool buildings. Roofs compose a large percentage of the absorptive surfaces in a city, estimated at 15–30 percent of the total impervious surface (Scott, 2006). The roofs of all of the structures in New York City, for example, account for 14 percent of the city's potentially absorptive surfaces, the lower end of the range. Green roofs also reduce the cooling cost of interior spaces of buildings by insulating them (Del Barrio, 1998). The insulating properties of green roofs reduce the emissions from summer cooling and winter heating that contribute to global warming. Reclaiming roof space with GI also simultaneously provides the myriad other health benefits associated with the water, air, food, biodiversity, and mental health services of GI (Frazer, 2005; Getter & Rowe, 2006).

Green roofs can also add an enormous amount of GI to a city while avoiding conflicts over the optimal use of finite urban space. Green roofs can therefore create some equity in who is receiving the benefits of reduced ambient heat. It is common that the ground-level greenspaces that reduce microclimatic heat are not evenly distributed among urban residents of varied economic status (Harlan, Brazel, Prashad, Stefanov, & Larsen, 2006; Solecki, Rosenzweig, & Parshall, 2005). Therefore, it is those with least individual capacity to reduce the risks from heat (with air conditioning) who also have the least amount of environmental support from GI. Those without GI in their neighborhood or working environments do not receive the microclimatic heat-reduction benefits of GI, but all structures, regardless of the amount of rent, have a roof.

Vector-borne and zoonotic diseases and climate[4]

The ecology of many infectious diseases is tightly linked to climate via impacts on the biology of pathogens, the arthropod vectors that transmit them, and the animal reservoirs that host them (Altizer, Ostfeld, Johnson, Kutz, & Harvell, 2013).

Figure 4.4: Green roof, Chicago

Source: Conservation Design Forum.

Changes in climate have already had significant impacts on many vector-borne and zoonotic diseases (Gage, Burkot, Eisen, & Hayes, 2008; Mills, Gage, & Khan, 2010; Rohr et al., 2011), and an increasing number of studies are trying to predict future climatic impacts on infectious disease risk (Altizer et al., 2013). There are three primary mechanisms through which climate change can affect vector-borne and zoonotic diseases: (1) geographic range shifts of vectors or reservoirs; (2) changes in rates of development, survival, and reproduction of vectors, reservoirs, and the pathogens that they carry; and (3) changes in biting rates of infected vectors or the prevalence of infection in reservoir or vector populations, which affects the likelihood of transmission resulting from contact with a human (Kovats, Campbell-Lendrum, McMichael, Woodward, & Cox, 2001; Mills et al., 2010; Reiter, 2001).

A meta-analysis of terrestrial species range shifts associated with climate change estimated that species are shifting to higher elevations at a rate of 11.0 meters per decade and higher latitudes at a rate of 16.9 kilometers per decade (Chen, Hill, Ohlemüller, Roy, & Thomas, 2011). Although expansion of pathogen and vector ranges to higher elevations and latitudes may be balanced by the contraction of their ranges in currently endemic areas (Lafferty, 2009), the introduction of pathogens into naive human populations with low acquired immunity or to areas with high human population density may have a significant public health impact (Dobson, 2009; Pascual & Bouma, 2009).

There are numerous examples of the impact of climate on the development, survival, and reproduction of vectors, reservoirs, and the pathogens they carry (Table 4.2). For example, higher temperatures have been associated with increased tick nymphal activity as a result of faster egg development after the birthing season (Gray, 2008). Mosquitoes need water to lay eggs and the availability of suitable sites is often affected by precipitation as well as the breeding requirements of each mosquito species, some of which prefer to breed in small pools while others prefer edges of larger, more stable water bodies (Gage et al., 2008). Mammalian pathogen reservoirs can be affected by high temperature directly through heat-related mortality as has occurred in the flying fox (*Pteropus* spp.) population in Australia (Welbergen, Klose, Markus, & Eby, 2008), or indirectly through increases in vegetative habitat and food resources related to increased temperature and precipitation (Yates et al., 2002). Many studies have shown that the extrinsic incubation period—the interval between the acquisition of an infectious agent by a vector and the time when a vector can transmit the infection—is highly temperature dependent (Noden, Kent, & Beier, 1995; Reisen, Meyer, Presser, & Hardy, 2000; Reisen, Fang, & Martinez, 2006; Watts, Burke, Harrison, Whitmire, & Nisalak, 1987). As the temperature increases, the rate of pathogen development inside the vector increases so that vectors can transmit the infection more quickly after a blood meal.

Increases in temperature can also increase the pathogen load within hosts via increased stress, which leads to a decrease in immune function (Mills et al., 2010). Increased temperature also increases the biting rates of arthropod vectors (Catalá, 1991; Patz et al., 1998; Semenza & Menne, 2009). Both increased host pathogen load and increased biting rates increase the likelihood of pathogen transmission to humans.

Although most climate and infectious disease research to date has focused on single-host, single-parasite interactions, climate change will also have profound effects on organisms and disease transmission at the ecosystem scale (Rohr et al., 2011). Rohr et al. (2011) use ecological theory to generate hypotheses about community-scale impacts of climate change on parasites and their hosts. For example, they suggest that there may be a greater probability of extinction of specialist parasites if their host species face extinction, while generalist parasites that can survive in a wider variety of hosts are more likely to adapt. Or similarly, they expect that parasites with multiple host lifecycles may be more likely to go extinct than those with direct transmission because of the high likelihood of at least one of their hosts going extinct or declining in number. As discussed in the next chapter, biodiversity has been shown to be a buffer for disease transmission in some systems due to the presence of "diluting hosts" that are incompetent reservoirs of pathogens (Ostfeld & Keesing, 2000). As we consider the impacts of climate change on the biology of pathogens, arthropod vectors, and animal reservoirs, we must also consider that climate-driven changes to host communities could alter disease risk (Rohr et al., 2011).

Finally, a key consideration for understanding the impacts of climate change on infectious disease is that the complexity of ecological systems makes it very difficult to predict whether climate change will increase or decrease vector-borne and zoonotic disease transmission around the globe as a whole. As described above, climate influences many aspects of disease ecology, and the direction of change in disease risk will be determined by the net effect of opposing interactions between transmission parameters (Rogers & Randolph, 2006). For example, increases in temperature are expected to increase biting rates and parasite and mosquito development, but if the temperature becomes too hot, mosquitoes will not be

Table 4.2: Selected examples of climatic factors influencing the transmission and distribution of vector-borne diseases

Disease (causative agent)	Vectors	Relevant climatic factors	Effects of climatic variability or climate change
Parasitic vector-borne diseases			
Malaria (*Plasmodium vivax*, *P. falciparum*)	Mosquitoes	Temperature, rainfall, humidity, El Niño-related effects, sea surface temperatures	Disease distribution; pathogen development in vector; development, reproduction, activity, distribution, and abundance of vectors; transmission patterns and intensity; outbreak occurrence
Leishmaniasis (*Leishmania spp.*)	Sand flies	Temperature, precipitation, El Niño-related effects	Disease incidence and outbreak occurrence; abundance, behavior, and distribution of vectors
Chagas disease (*Trypanosoma cruzi*)	Triatomine bugs	Temperature, precipitation, humidity, severe weather event	Vector distribution, increased infestation of houses by vector
Onchocerciasis (*Onchocerca volvulus*)	Black flies	Temperature	Transmission intensity
Arboviruses			
Dengue fever (Dengue virus)	Mosquitoes	Temperature, precipitation	Outbreaks, mosquito breeding, abundance, transmission intensity (extrinsic incubation period)
Yellow fever (Yellow fever virus)	Mosquitoes	Temperature, precipitation	Outbreaks, incidence; distribution, abundance, and breeding of mosquitoes, transmission intensity (extrinsic incubation period)
Chikungunya fever (Chikungunya virus)	Mosquitoes	Temperature, precipitation	Outbreaks, vector breeding and abundance, transmission intensity (extrinsic incubation period)
West Nile virus disease (West Nile virus)	Mosquitoes	Temperature, precipitation	Transmission rates, pathogen development in vector, distribution of disease and vector

Disease	Vector	Climate variables	Effects
Rift Valley fever (*Rift Valley Fever virus*)	Mosquitoes	Precipitation, sea surface temperatures	Outbreaks; vector breeding and abundance, transmission intensity (extrinsic incubation period)
Ross River virus disease (*Ross River Virus*)	Mosquitoes	Temperature, precipitation, sea surface temperatures	Outbreaks; vector breeding and abundance, transmission intensity (extrinsic incubation period)
Tick-borne encephalitis (Tick-borne Encephalitis virus)	Ticks	Temperature, precipitation, humidity	Vector distribution, phenology of host-seeking by vector
Bacterial and rickettsial diseases			
Lyme borreliosis (*Borrelia burgdorferi, B. garinii, B. afzelii*, or other related *Borrelia*)	Ticks	Temperature, precipitation	Frequency of cases, phenology of host-seeking by vector, vector distribution
Tularemia (*Francisella tularensis*)	Ticks	Temperature, precipitation	Case frequency and onset
Human granulocytic anaplasmosis (*Anaplasma phagocytophilum*)	Ticks	Temperature, precipitation	Vector distribution, phenology of host-seeking by vector
Human monocytic ehrlichiosis (*Ehrlichia chafeensis*)	Ticks	Temperature, precipitation	Phenology of host-seeking by vector
Plague (*Yersinia pestis*)	Fleas	Temperature, precipitation, humidity, El Niño–related events	Development and maintenance of pathogen in vector; survival and reproduction of vectors and hosts; occurrence of historical pandemics and regional outbreaks; distribution of disease

Source: Gage et al. (2008).

able to survive (Rohr et al., 2011). The overall impact on malaria transmission, for example, depends on both of these parameters (Rogers & Randolph, 2006) as well as human adaptation measures. Although scientists recognize the challenge of effective forecasting of infectious disease impacts due to climate change, there is broad consensus that interdisciplinary research to improve our understanding of mechanistic factors related to disease systems, long-term ecological monitoring, and consideration of human adaptation and control measures are essential elements for predicting future disease occurrence (Altizer et al., 2013; Ostfeld, 2009; Paull & Johnson, 2013; Rohr et al., 2011).

Summary

The environmental changes brought about by climate change will be accompanied by a host of public health challenges that GI can play a role in ameliorating. Conserving and expanding GI and greening our gray infrastructure can help mitigate and adapt to climate change. Mitigation is done by capturing the atmospheric carbon that contributes to global warming and more severe climatic variability. Using GI to adapt to climate change is beneficial to health through its ability to lessen the effects of extreme weather events and urban heat islands. GI cannot make a significant dent in the levels of carbon currently being emitted into the atmosphere, but GI does help and, at the same time, provides the co-benefits explored in previous and subsequent chapters. These benefits include mitigating the proliferation of infectious diseases and helping us to adapt to the new set of risks that are inevitably coming.

Notes

1. The outcomes measured in this analysis were: direct impacts of heat and cold (cardiovascular disease deaths); food and water-borne disease (diarrhea episodes); vector-borne disease (malaria and dengue cases); natural disasters (fatal unintentional injuries); and risk of malnutrition (non-availability of recommended daily calorie intake).
2. A beautiful set of interactive maps shows this distribution in an easily accessible format: http://coolclimate.berkeley.edu/maps
3. Climate change will also continue to alter blue infrastructure and freshwater and marine ecosystems.
4. Section contributed by Micah Hahn.

References

Akbari, H. (2005). Energy saving potentials and air quality benefits of urban heat island mitigation (working paper). Berkeley, CA: Lawrence Berkeley National Laboratory.

Akbari, H., Kurn, D., Bretz, S., & Hanford, J. (1997). Peak power and cooling energy savings of shade trees. *Energy and Buildings*, *25*(2), 139–48.

Akbari, H., Pomerantz, M., & Taha, H. (2001). Cool surfaces and shade trees to reduce energy use and improve air quality in urban areas. *Solar Energy*, *70*(3), 295–310.

Allen, A., & Segal-Gidan, F. (2007). Heat-related illness in the elderly. *Clinical Geriatrics*, *15*(7), 37–45.

Altizer, S., Ostfeld, R. S., Johnson, P. T. J., Kutz, S., & Harvell, C. D. (2013). Climate change and infectious diseases: From evidence to a predictive framework. *Science*, *341*(6145), 514–19.

Amati, M., & Taylor, L. (2010). From green belts to green infrastructure. *Planning Practice and Research*, *25*(2), 143–55.

Arnfield, A. J. (2003). Two decades of urban climate research: A review of turbulence, exchanges of energy and water, and the urban heat island. *International Journal of Climatology*, *23*(1), 1–26.

Barbier, E. B., Koch, E. W., Silliman, B. R., Hacker, S. D., Wolanski, E., Primavera, J., et al. (2008). Coastal ecosystem-based management with nonlinear ecological functions and values. *Science*, *319*(5861), 321–3.

Bellassen, V., & Luyssaert, S. (2014). Carbon sequestration: Managing forests in uncertain times. *Nature*, *506*(7487), 153–5.

Borden, K. A., & Cutter, S. L. (2008). Spatial patterns of natural hazards mortality in the United States. *International Journal of Health Geographics*, *7*, 64.

Brack, C. L. (2002). Pollution mitigation and carbon sequestration by an urban forest. *Environmental Pollution*, *116*, S195–S200.

Campbell-Lendrum, D. H., Corvalán, C. F., & Prüss-Ustün, A. (2003). How much disease could climate change cause? In A. J. McMichael, D. Campbell-Lendrum, C. Corvalan, K. Ebi, A. Githeko, J. Scheraga, & A. Woodward (Eds.), *Climate change and human health: Risks and responses* (pp. 133–58). Geneva: World Health Organization.

Catalá, S. (1991). The biting rate of Triatoma infestans in Argentina. *Medical and Veterinary Entomology*, *5*(3), 325–33.

CDC. (2009). Extreme heat prevention guide. Retrieved May 29, 2015, from www.bt.cdc.gov/disasters/extremeheat/heat_guide.asp

Chen, I.-C., Hill, J. K., Ohlemüller, R., Roy, D. B., & Thomas, C. D. (2011). Rapid range shifts of species associated with high levels of climate warming. *Science, 333*(6045), 1024–6.

Cockburn, A., St. Clair, J., & Silverstein, K. (1999). The politics of "natural" disaster: Who made Mitch so bad? *International Journal of Health Services, 29*(2), 459–62.

Confalonieri, U., Menne, B., Akhtar, R., Ebi, K., Hauengue, M., Kovats, R., et al. (2007). Human health. In M. Parry, O. Canziani, J. Palutikof, P. van der Linden, & C. Hanson (Eds.), *Climate change 2007: Impacts, adaptation and vulnerability. Contribution of working group II to the fourth assessment report of the intergovernmental panel on climate change* (pp. 391–431). Cambridge: Cambridge University Press.

Costello, A., Abbas, M., Allen, A., Ball, S., Bell, S., Bellamy, R., et al. (2009). Managing the health effects of climate change. *The Lancet, 373*(9676), 1693–733.

Coutts, C., & Berke, T. (2013). The extent and context of human health considerations in London's spatial development and climate action strategy. *Journal of Urban Planning and Development, 139*(4), 322–30.

Del Barrio, E. P. (1998). Analysis of the green roofs cooling potential in buildings. *Energy and Buildings, 27*(2), 179–93.

Dobson, A. (2009). Climate variability, global change, immunity, and the dynamics of infectious diseases. *Ecology, 90*(4), 920–7.

Ebi, K. L. (2010). *The ABCD of adaptation.* Washington, DC: American Meteorological Society.

English, P. B., Sinclair, A. H., Ross, Z., Anderson, H., Boothe, V., Davis, C., et al. (2009). Environmental health indicators of climate change for the United States: Findings from the State Environmental Health Indicator Collaborative. *Environmental Health Perspectives, 117*(11), 1673–81.

Frazer, L. (2005). Paving paradise: The peril of impervious surfaces. *Environmental Health Perspectives, 13*(7), 456–2.

Frumkin, H., Hess, J., Luber, G., Malilay, J., & McGeehin, M. (2008). Climate change: The public health response. *American Journal of Public Health, 98*(3), 435–45.

Gage, K. L., Burkot, T. R., Eisen, R. J., & Hayes, E. B. (2008). Climate and vector-borne diseases. *American Journal of Preventive Medicine, 35*(5), 436–50.

Getter, K., & Rowe, D. (2006). The role of extensive green roofs in sustainable development. *HortScience, 41*(5), 1276–85.

Gill, S., Handley, J., Ennos, A., & Pauleit, S. (2007). Adapting cities for climate change: The role of the green infrastructure. *Built Environment, 33*(1), 115–33.

Goklany, I. M. (2007). *Death and death rates due to extreme weather events: Global and U.S. trends, 1900–2006.* London: International Policy Network.

Gray, J. S. (2008). Ixodes ricinus seasonal activity: Implications of global warming indicated by revisiting tick and weather data. *International Journal of Medical Microbiology, 298*(S1), 19–24.

Hajat, S., Vardoulakis, S., Heaviside, C., & Eggen, B. (2014). Climate change effects on human health: Projections of temperature-related mortality for the UK during the 2020s, 2050s and 2080s. *Journal of Epidemiology and Community Health, 68,* 641–8.

Harlan, S. L., Brazel, A. J., Prashad, L., Stefanov, W. L., & Larsen, L. (2006). Neighborhood microclimates and vulnerability to heat stress. *Social Science & Medicine, 63*(11), 2847–63.

Heisler, G., Grant, R., Grimmond, S., & Souch, C. (1995). Urban forests—cooling our communities? In C. Kollin & M. Barratt (Eds.), *Proceedings: 7th national urban forest conference* (pp. 31–4). Washington, DC: American Forests.

Hulme, M., Jenkins, G., Lu, X., Turnpenny, J., Mitchell, T., Jones, R., et al. (2002). *Climate change scenarios for the United Kingdom: The UKCIP02 scientific report.* Norwich, UK: University of East Anglia.

Jones, C., & Kammen, D. (2014). Spatial distribution of US household carbon footprints reveals suburbanization undermines GHG benefits of urban population density. *Environmental Science & Technology, 48*(2), 895–902.

Kalkstein, L. S., & Greene, J. S. (1997). An evaluation of climate/mortality relationships in large U.S. cities and the possible impacts of a climate change. *Environmental Health Perspectives, 105*(1), 84–93.

Karl, T., Melillo, J., & Peterson, T. (Eds.). (2009). *Global climate change impacts in the United States.* New York, NY: Cambridge University Press.

Kistin, E. J., Fogarty, J., Pokrasso, R. S., McCally, M., & McCornick, P. G. (2010). Climate change, water resources and child health. *Archives of Disease in Childhood, 95*(7), 545–9.

Körner, C. (2000). Biosphere responses to CO_2 enrichment. *Ecological Applications, 10*(6), 1590–619.

Kosatsky, T. (2005). The 2003 European heat wave. *Eurosurveillance, 10*(7–9), 148–9.

Kovats, R. S., Campbell-Lendrum, D. H., McMichael, A. J., Woodward, A., & Cox, J. S. (2001). Early effects of climate change: Do they include changes in vector-borne disease? *Philosophical Transactions of the Royal Society of London—Series B: Biological Sciences, 356*(1411), 1057–68.

Lafferty, K. D. (2009). The ecology of climate change and infectious disease. *Ecology, 90*(4), 888–900.

Lemmen, D. S., & Warren, F. J. (2004). *Climate change impacts and adaptation: A Canadian perspective*. Ottawa: Climate Change Impacts and Adaptation Directorate, Natural Resources Directorate.

Lichtman, J. H., Leifheit-Limson, E. C., & Goldstein, L. B. (2014). Weather changes may be linked with stroke hospitalization, death. Retrieved April 17, 2014, from http://newsroom.heart.org/news/weather-changes-may-be-linked-with-stroke-hospitalization-death

Lukes, R., & Kloss, C. (2008). *Managing wet weather with green infrastructure. Municipal handbook, green streets.* Washington, DC: US Environmental Protection Agency.

McGeehin, M. A., & Mirabelli, M. (2001). The potential impacts of climate variability and change on temperature-related morbidity and mortality in the United States. *Environmental Health Perspectives, 109 Suppl* (May), 185–9.

McMichael, A. J. (2012). Insights from past millennia into climatic impacts on human health and survival. *Proceedings of the National Academy of Sciences, 109*(13), 4730–7.

McMichael, A. J. (2013). Globalization, climate change, and human health. *New England Journal of Medicine, 368*(14), 1335–43.

McMichael, A. J., Campbell-Lendrum, D., Corvalan, C., Ebi, K., Githeko, A., Scheraga, J., & Woodward, A. (2003). *Climate change and human health: Risks and responses.* Geneva: WHO.

McMichael, A. J., Campbell-Lendrum, D., Kovats, S., Edwards, S., Wilkinson, P., Wilson, T., et al. (2004). Global climate change. In M. Ezzati, A. Lopez, A. Rodgers, & C. Murray (Eds.), *Comparative quantification of health risks: Global and regional burden of disease due to selected major risk factors* (pp. 1543–649). Geneva: WHO.

McMichael, A. J., Woodruff, R., & Hales, S. (2006). Climate change and human health: Present and future risks. *The Lancet, 367*(9513), 859–69.

Millennium Ecosystem Assessment. (2005). *Ecosystems and human well-being: Current state and trends.* Washington, DC: Island Press.

Mills, J. N., Gage, K. L., & Khan, A. S. (2010). Potential influence of climate change on vector-borne and zoonotic diseases: A review and proposed research plan. *Environmental Health Perspectives, 118*(11), 1507–14.

Munich, R. E. (2007). *Natural catastrophes 2007: Analyses, assessments, positions.* München, Germany: Munich RE.

Munich, R. E. (2013). *Natural catastrophes 2012: Analyses, assessments, positions.* München, Germany: Munich RE.

National Institute of Environmental Health Sciences. (2010). *A human health perspective on climate change.* Research Triangle Park, NC: National Institute of Environmental Health Sciences.

Nobre, C. A., Sellers, P. J., & Shukla, J. (1991). Amazonian deforestation and regional climate change. *Journal of Climate, 4*(10), 957–88.

Noden, B. H., Kent, M. D., & Beier, J. C. (1995). The impact of variations in temperature on early Plasmodium falciparum development in Anopheles stephensi. *Parasitology, 111,* 539–45.

Nowak, D. J. (1993). Atmospheric carbon reduction by urban trees. *Journal of Environmental Management, 37,* 207–17.

Nowak, D. J., & Crane, D. E. (2002). Carbon storage and sequestration by urban trees in the USA. *Environmental Pollution, 116*(3), 381–9.

Oke, T. R. (1982). The energetic basis of the urban heat island. *Quarterly Journal of the Royal Meteorological Society, 108*(455), 1–24.

Ostfeld, R. S. (2009). Climate change and the distribution and intensity of infectious diseases. *Ecology, 90*(4), 903–5.

Ostfeld, R. S., & Keesing, F. (2000). Biodiversity and disease risk: The case of lyme disease. *Conservation Biology, 14*(3), 722–8.

Pascual, M., & Bouma, M. J. (2009). Do rising temperatures matter? *Ecology, 90*(4), 906–12.

Patz, J. A., Strzepek, K., Lele, S., Hedden, M., Greene, S., Noden, B., et al. (1998). Predicting key malaria transmission factors, biting and entomological inoculation rates, using modelled soil moisture in Kenya. *Tropical Medicine and International Health, 3*(10), 818–27.

Patz, J. A., Campbell-Lendrum, D., Holloway, T., & Foley, J. A. (2005). Impact of regional climate change on human health. *Nature, 438*(7066), 310–17.

Pauleit, S., & Duhme, F. (2000a). Assessing the environmental performance of land cover types for urban planning. *Landscape and Urban Planning, 52*(1), 1–20.

Pauleit, S., & Duhme, F. (2000b). GIS assessment of Munich's urban forest structure for urban planning. *Journal of Arboriculture, 26*(1), 133–41.

Paull, S. H., & Johnson, P. T. (2013). Can we predict climate-driven changes to disease dynamics? Applications for theory and management in the face of uncertainty. In J. Brodie, E. Post, & D. Doak (Eds.), *Wildlife conservation in a changing climate* (pp. 109–28). Chicago, IL: University of Chicago Press.

Reisen, W. K., Meyer, R. P., Presser, S. B., & Hardy, J. L. (2000). Effect of temperature on the transmission of western equine encephalomyelitis and St. Louis encephalitis viruses by Culex tarsalis (Diptera: Culicidae). *Journal of Medical Entomology, 30*(1), 151–60.

Reisen, W. K., Fang, Y., & Martinez, V. M. (2006). Effects of temperature on the transmission of West Nile virus by Culex tarsalis (Diptera: Culicidae). *Journal of Medical Entomology, 43*(2), 309–17.

Reiter, P. (2001). Climate change and mosquito-borne disease. *Environmental Health Perspectives, 109*(Suppl 1), 141–61.

Rogers, D. J., & Randolph, S. E. (2006). Climate change and vector-borne diseases. *Advances in Parasitology, 62*(5), 345–81.

Rohr, J. R., Dobson, A. P., Johnson, P. T. J., Kilpatrick, A. M., Paull, S. H., Raffel, T. R., et al. (2011). Frontiers in climate change-disease research. *Trends in Ecology & Evolution, 26*(6), 270–7.

Scott, M. (2006). Beating the heat in the world's big cities. Retrieved April 11, 2013, from http://earthobservatory.nasa.gov/Features/GreenRoof/greenroof2.php

Semenza, J. C., & Menne, B. (2009). Climate change and infectious diseases in Europe. *The Lancet Infectious Diseases, 9*(6), 365–75.

Sharp, J. D., Jaccard, M. K., & Keith, D. W. (2009). Anticipating public attitudes toward underground CO_2 storage. *International Journal of Greenhouse Gas Control, 3*(5), 641–51.

Sheffield, P. E., & Landrigan, P. J. (2011). Global climate change and children's health: Threats and strategies for prevention. *Environmental Health Perspectives, 119*(3), 291–8.

Smith, C., & Levermore, G. (2008). Designing urban spaces and buildings to improve sustainability and quality of life in a warmer world. *Energy Policy, 36*(12), 4558–62.

Solecki, W., Rosenzweig, C., & Parshall, L. (2005). Mitigation of the heat island effect in urban New Jersey. *Environmental Hazards, 6*, 39–49.

Souch, C., & Grimmond, S. (2006). Applied climatology: Urban climate. *Progress in Physical Geography, 30*(2), 270–9.

Spronken-Smith, R., & Oke, T. (1999). Scale modelling of nocturnal cooling in urban parks. *Boundary-Layer Meteorology, 93*, 287–312.

Tyrväinen, L., Pauleit, S., Seeland, K., & de Vries, S. (2005). Benefits and uses of urban forests and trees. In C. Konijnendijk, K. Nilsson, T. Randrup, & J. Schipperijn (Eds.), *Urban forests and trees: A reference book* (pp. 81–114). Berlin: Springer.

US EPA. (2003). *Cooling summertime temperatures: Strategies to reduce urban heat islands.* Washington, DC: US EPA.

US EPA. (2013a). Heat island effect. Retrieved April 11, 2013, from www.epa. gov/hiri/

US EPA. (2013b). Heat island effect: Heat island impacts. Retrieved April 11, 2013, from www.epa.gov/heatisland/impacts/index.htm

US EPA. (2013c). Inventory of U.S. greenhouse gas emissions and sinks: 1990–2011. *Journal of the ICRU, 9*(2), ES1–ES26.

Watkiss, P., Horrocks, L., Pye, S., Searl, A., & Hunt, A. (2009). *Impacts of climate change in human health in Europe. PESETA–Human health study.* Luxembourg: European Commission Joint Research Center.

Watts, D. M., Burke, D. S., Harrison, B. A., Whitmire, R. E., & Nisalak, A. (1987). Effect of temperature on the vector efficiency of Aedes aegypti for dengue 2 virus. *American Journal of Tropical Medicine and Hygiene, 36*(1), 143–52.

Welbergen, J. A., Klose, S. M., Markus, N., & Eby, P. (2008). Climate change and the effects of temperature extremes on Australian flying-foxes. *Proceedings of the Royal Society B: Biological Sciences, 275*(1633), 419–25.

Wilby, R. (2003). Past and projected trends in London's urban heat island. *Weather, 58*, 251–60.

Wolf, K. (2004). *Trees, parking and green law: Strategies for sustainability.* Stone Mountain, GA: Georgia Forestry Commission, Urban and Community Forestry.

Woodbury, P. B., Smith, J. E., & Heath, L. S. (2007). Carbon sequestration in the U.S. forest sector from 1990 to 2010. *Forest Ecology and Management, 241*(1–3), 14–27.

World Meteorological Organization. (2014). *WMO statement on the status of the global climate in 2013.* Geneva: World Meteorological Organization.

Yates, T. L., Mills, J. N., Parmenter, C. A., Ksiazek, T. G., Parmenter, R. R., Vande Castle, J. R., et al. (2002). The ecology and evolutionary history of an emergent disease: Hantavirus pulmonary syndrome. *BioScience, 52*(11), 989–98.

Chapter Five

Infectious disease ecology

Micah Hahn

Infectious disease ecology is a rapidly evolving field focused on understanding how hosts, pathogens, vectors, and their environment evolve, respond, and interact with one another in ways that influence the spread of disease (Ostfeld, Keesing, & Eviner, 2008). A recent review of the land-use change and infectious disease ecology literature identified over 300 articles, the majority of which documented increased pathogen transmission with anthropogenic alterations to the landscape (Gottdenker, Streicker, Faust, & Carroll, 2014). Landscape epidemiology is a field nestled within the broad study of disease ecology that explicitly examines the influence of landscape structure on disease risk (Ostfeld, Glass, & Keesing, 2005). The landscape and GI is increasingly recognized as a barrier or conduit of disease amplification and spread in human, domestic animal, and wildlife populations. Disease ecology and its complementary fields all appreciate that the consequences of altering GI is part of a dynamic process of feedbacks and cascading impacts of ecosystem perturbation that may not be evident for several years (Defries, Foley, & Asner, 2004).

A number of studies from these fields have demonstrated the various pathways through which patterns of GI can mediate infectious disease spread. The composition and configuration of GI can influence disease risk, for example, by determining the amount of viable habitat and therefore the size of the zoonotic reservoir (long-term host of a pathogen) and vector populations that maintain and transmit pathogens to humans. Alternatively, impacts may be more indirect if alterations to the landscape shift the biodiversity of an ecosystem in ways that limit or propagate pathogen spread within the reservoir community, a primary determinant of human risk. Ultimately, however, cultural, social, and political forces shape these human–environment interactions. Therefore, understanding the human dimensions of landscape change is vital for understanding infectious disease ecology.

Zoonotic disease

Nearly two-thirds of human infections are zoonotic (Karesh et al., 2012), meaning that they are animal pathogens that have been transmitted to humans. Studies from a variety of ecological contexts have documented how landscape changes can influence the frequency and intimacy of interaction between humans and zoonotic disease reservoirs and propagate microbial transmission between species.

One example of these interactions is the trade in wild animal meat. Hunting and sale of wild animals, or bushmeat, is an important source of income and protein for many rural communities in the tropics, with estimates exceeding five million tons of meat extracted annually (Nasi, Taber, & Van Vliet, 2011). Intimate blood and bodily fluid contact between the hunters and the hunted create an effective interface for the introduction of novel, zoonotic infectious agents into the human population (LeBreton et al., 2012).

A number of social, cultural, and economic factors are driving the expansion of the bushmeat trade. One of these is the commercial logging industry that has transformed hunting from a subsistence activity to a lucrative commercial under-taking (Figure 5.1) (Karesh & Noble, 2009; Poulsen, Clark, Mavah, & Elkan, 2009). In addition to the increased demand for bushmeat from workers in the logging industry and their families, roads built by logging companies dramatically increase access to remote forest areas and link these hunting areas to markets, thereby significantly reducing the cost of business for bushmeat hunters. In Central Africa, a hub of the "bushmeat crisis" (Bennett et al., 2006; Wolfe, Daszak, Kilpatrick, & Burke, 2005), approximately 38 percent of all roads were built for logging activities (Laporte, Stabach, Grosch, Lin, & Goetz, 2007).

For some diseases, such as Ebola, monkeypox, or anthrax, a single transmission event between a wild animal and a human can cause a localized outbreak (Wolfe et al., 2005). For other pathogens, there is evidence of repeated spillover events, where the virus jumps from its natural animal reservoir to a human with little or

Figure 5.1: Logging in the Central African Republic
Source: JG Collomb.

no subsequent human-to-human transmission. During these repeated incursions (a phenomenon known as "viral chatter"), animal viruses may acquire mutations that make them more likely to spread among humans, as was the case with the repeated introductions of Simian Immunodeficiency Virus (SIV) that led to the global emergence of HIV (LeBreton et al., 2012; Wolfe et al., 2005). The growing bushmeat trade, augmented by the expansion of the logging road network, has set the stage for these types of repeated contact events between reservoirs and humans.

In addition to increasing the frequency of direct contact between humans and zoonotic disease reservoirs, the management and alteration of landscape can influence indirect interactions among people, wildlife, and domestic animals through utilization of shared resources. In Kibale, Uganda, a long-term ecological study has documented the presence of forest fragments leftover on land unfit for agriculture near the border of Kibale National Park that are areas of intense human–primate interaction (Goldberg, Paige, & Chapman, 2012). Not only are these fragments home to dense primate populations, they are also important sources of fuel wood, timber, medicinal plants, and materials for household use for people living near the park. The primates living in the fragments likewise venture into the human settlements surrounding the fragments in search of food, coming into contact with livestock and their feces along the way. In order to deter crop raiding by primates, farmers apply a mixture of sand and cattle dung to ears of maize along the edge of their fields near forest fragments. Additionally, humans, primates, and domestic animals often have common open water sources. The result of these overlapping living spaces and resources is increased bacterial transmission among species as evidenced by the genetic similarity of gut bacteria in primates, humans, and livestock living near the same forest fragment (Goldberg, Gillespie, Rwego, Estoff, & Chapman, 2008). Moreover, the degree of genetic similarity of these bacteria parallels the relative degree of anthropogenic disturbance in the fragments (based on measures of encroachment, forest clearing rates, and intensity of human use) (Goldberg et al., 2008).

Similar findings emerged from a study of landscape composition and configuration and Nipah virus (NiV) in Bangladesh (Hahn et al., 2014). NiV is harbored by *Pteropus* fruit bats and is transmitted to humans primarily through the consumption of raw date palm sap that has been contaminated by the saliva or urine of infected bats that drank from the collection pots hung high on the trunk of tapped trees (Luby et al., 2006). Since the documentation of NiV in Bangladesh in 2001, human cases have occurred within a fairly well-defined band in the Northwest and central part of the country, known colloquially as the Nipah Belt (Luby, Gurley, & Hossain, 2009). Disease ecology studies in the region found that the Nipah Belt has a higher human population density and more dense patchwork of smaller forest remnants than the rest of the country (Hahn et al., 2014). Additionally, the forest patches in the Nipah Belt are more biologically diverse, likely because the communities living in the region have planted them as a source of fruit (Hahn et al., 2014). As a result, *Pteropus* bats roosting in the remaining forest patches in the Nipah Belt are essentially living in people's backyards and eating from their homegardens. Again, the management of the landscape to meet human needs created a situation where shared habitat and resources increased the risk of disease transmission between a zoonotic reservoir and human communities. Although a tempting easy-fix might be to cut down the forest in high risk areas to prevent human–animal interactions, this short-sighted solution would likely have devastating repercussions not only for the primate and bat communities living in these areas, but also for the human populations who depend on these natural resources for food and livelihoods.

Vector-borne disease

Vector-borne diseases are transmitted to humans by arthropods, such as mosquitoes, ticks, fleas, and a number of other blood-sucking insects. Almost 30 percent of emerging infections in the last decade were vector-borne diseases and this number has risen since the 1940s (Jones et al., 2008). Changes to the landscape can affect vector breeding sites and the microclimate in ways that significantly impact the rate of larval development, biting frequency, and survival of the vector (Patz, Olson, Uejio, & Gibbs, 2008; Reiter, 2001).

The global demand for cropland and pasture drove the expansion of agricultural land in the last half of the twentieth century, primarily at the expense of intact tropical forests (Gibbs et al., 2010). The impact of deforestation in the tropics on vector-borne disease risk has been shown across continents using fine-scale field studies as well as more coarse-scale remote sensing studies used to assess large geographic extents (Walsh, Molyneux, & Birley, 1993). For example, in western Kenya, Afrane, Lawson, Githeko and Yan (2005) showed that the average ambient temperature in deforested areas was 0.5°C warmer than forested areas. As a result, the reproductive cycle of the *Anopheles* mosquitoes that transmit malaria living in the deforested areas was accelerated by almost three days, resulting in an increased biting frequency and risk of malaria transmission to humans (Afrane et al., 2005). In Costa Rica, cutaneous leishmaniasis transmission is augmented in areas where social marginalization has led to forest destruction for agriculture (Chaves, Cohen, Pascual, & Wilson, 2008). Deforestation in this context alters leishmaniasis epidemiology through multiple pathways including impacts on the biodiversity of the rodent population that are the reservoirs of the leishmania parasite, changes to the habitat of the sand fly vector, and small-scale impacts on climate variability in the region that influence the biology of both the sand fly and the leishmania parasite (Chaves et al., 2008). In West Africa, remote sensing imagery was used to assess the conversion of forests into savanna and urban areas between 1973 and 1990 (Wilson et al., 2002). Coinciding with this deforestation, the proportion of the black fly vector that transmits the more severe form of onchocerciasis increased, likely due to the increase in available breeding sites (Wilson et al., 2002).

Water management projects and the subsequent changes to the landscape can also impact vector-borne disease risk by providing new breeding sites both during construction of water projects and through the long-term ecosystems shifts they propagate (Patz, Graczyk, Geller, & Vittor, 2000). For example, on the Front

Range of Colorado, irrigated agriculture has been shown to be an important risk factor for West Nile virus (WNV) transmission, likely because these irrigated areas provide breeding sites for *Culex tarsalis*, the primary WNV mosquito vector in the region (Eisen et al., 2010). Japanese encephalitis (JE) is a mosquito-borne disease that has a range of clinical manifestations from mild, flu-like symptoms to seizures, encephalitis, and life-long neurological disease. Primarily transmitted in Southeast Asia and the Western Pacific, there are close ties between JE epidemiology and the expansion of irrigated rice fields that offer breeding sites for JE vector populations (Keiser et al., 2005). Schistosomiasis is an intravascular disease caused by parasitic trematode worms that infect both mammals and freshwater snails to complete their life cycle. Extensive research has demonstrated the link between dam construction and schistosomiasis risk in countries throughout the world (Steinmann, Keiser, Bos, Tanner, & Utzinger, 2006). In China, where over 50 million people are at risk of infection, much attention has been given to the impact of the Three Gorges Dam on schistosomiasis transmission (McManus et al., 2010; Zhu, Xiang, Yang, Wu, & Zhou, 2008). Upstream lakes behind the dam created snail breeding sites along the river banks, and seasonal downstream flows were shifted in ways that expanded marshland, changed silt deposition, and prevented natural flooding that once served to modulate the snail population (McManus et al., 2010).

Biodiversity and disease risk

The dilution effect is a term used to describe the observation that high host diversity in an ecosystem can reduce the risk of human disease (Ostfeld & Keesing, 2000). This observation was first established for Lyme disease, which has since become the host–parasite system most often used to exemplify the concept. Lyme disease is caused by the *Borrelia burgdorferi* bacterium that is transmitted to humans by the bite of the black-legged tick of the genus *Ixodes* (Figure 5.2). Ixodid ticks go through four life stages (egg, larva, nymph, and adult) and take a single blood meal at each of the last three. The larval ticks are generally born uninfected with *B. burgdorferi* but then acquire the parasite by feeding on an infected vertebrate host (Ostfeld & Keesing, 2000). Larvae and nymphs are highly non-specific in terms of the vertebrate hosts they feed on; however, the hosts vary significantly in the probability of transmitting *B. burgdorferi* to a feeding tick. For example, the white-footed mouse (*Peromyscus leucopus*) is a highly competent reservoir for *B. burgdorferi* because 40–80 percent of larval ticks that feed on an infected white-footed mouse acquire the pathogen (Mather, 1993). Other animals such as raccoons, squirrels, and opossums are considered incompetent reservoirs because they do not pass on the bacterium as readily to a feeding tick.

Based on this variation in transmission efficiency among reservoirs, the dilution effect posits that the infection prevalence of the tick community is related to the species composition of the vertebrate reservoir community. In diverse communities where there are a number of different vertebrates for feeding ticks, the probability of a tick becoming infected through a blood meal is lower than in a species-poor community where ticks are feeding predominately on the highly competent white-footed mouse reservoir (Ostfeld & Keesing, 2000). In essence, the incompetent reservoirs are "diluting" the impact of the white-footed mouse on increasing tick infection prevalence and subsequent risk of human infection.

In order to tie the dilution effect to landscape, we have to understand how the white-footed mouse population responds to landscape patterns. Nupp and Swihart (1996) studied the connections between forest patch size and white-footed mouse populations, and they found that the density of the mouse population was highest

Figure 5.2: Black-legged tick
Source: James Gathany.

in study sites with less forest cover. White-footed mice are habitat and dietary generalists, meaning that they can thrive in a variety of environments. In relation to Lyme disease risk, this means that in fragmented forest landscapes, white-footed mice are found in higher proportions than larger vertebrates (that are also less competent *B. burgdorferi* reservoirs) that cannot survive in smaller patches (LoGiudice, Ostfeld, Schmidt, & Keesing, 2003) and, as a result, human disease risk near these patches is increased.

Although the dilution effect has been demonstrated in a variety of disease systems ranging from Sin Nombre virus in the Western United States to WNV in the Midwest (Salkeld, Padgett, & Jones, 2013), the relationship between biodiversity and disease risk is highly dependent on the ecological context of a disease system (Salkeld et al., 2013). For example, in a field study in East Africa, Young et al. (2014) showed that declines of large wild herbivores had no impact on the diversity of the rodent population as the dilution effect hypothesizes, but instead increased human disease risk through a doubling of the rodent population and associated flea populations that transmit the *Bartonella* bacteria. This study demonstrates the importance of considering the local ecology of a disease system in order to understand the impacts of landscape on reservoir populations and human disease risk.

Social context of landscape change

Although the cases presented in this section rely heavily on quantitative epidemiological and ecological research, it is essential to recognize and elucidate the cultural, political, and economic drivers of land use and landscape change in order to adapt this landscape-disease knowledge into effective public health interventions.

If we scan back through the disease systems presented in this section, a common theme is that the emergence or spread of each pathogen was the result of anthropogenic land-use change driven by social or economic forces. In Kibale and Costa Rica, the need for agricultural land caused the conversion of forest to cropland. In Central Africa and Bangladesh, harvesting of natural resources as a livelihood brought humans into close contact with disease reservoirs. Further inquiry into these social processes reveals that *who* is involved in these land-use practices is also a product of social class, culture, and politics. It is tempting, for example, to ban bushmeat hunting, date palm sap collection, and agricultural expansion; however, a more holistic understanding of these practices shows that these policies would eliminate a vital protein source, a livelihood, and a cultural practice that has been passed down through generations. Measurement of the ecological interactions between humans, domestic and wild animals, and the ecosystems we share can provide vital information for understanding the transmission of pathogens between species. However, the cultural and social beliefs and behaviors of people determine how, when, and where these interactions, and ultimately zoonotic and vector-borne disease transmission, manifest. An appreciation for the human dimension of land use and infectious disease is therefore fundamental for developing interventions to protect the ecosystems that support health.

Summary

In summary, landscape patterns impact human disease risk through impacts on the vector and reservoir populations and by influencing interactions between these animals and humans. The ecological mechanisms that determine risk are complex and are often a series of cascading impacts that may be apparent immediately or play out over months to years. Long-term ecological studies that address the social and cultural context of landscape alteration are our best tool for understanding these processes.

References

Afrane, Y., Lawson, B. W., Githeko, A. K., & Yan, G. (2005). Effects of microclimatic changes caused by land use and land cover on duration of gonotrophic cycles of Anopheles gambiae (Diptera: Culicidae) in Western Kenya Highlands. *Journal of Medical Entomology, 42*(6), 974–80.

Bennett, E. L., Blencowe, E., Brandon, K., Brown, D., Burn, R. W., Cowlishaw, G., et al. (2006). Hunting for consensus: Reconciling bushmeat harvest, conservation, and development policy in West and Central Africa. *Conservation Biology, 21*(3), 884–7.

Chaves, L. F., Cohen, J. M., Pascual, M., & Wilson, M. L. (2008). Social exclusion modifies climate and deforestation impacts on a vector-borne disease. *PLoS Neglected Tropical Diseases, 2*(1), e176.

Defries, R. S., Foley, J., & Asner, G. (2004). Land-use choices: Balancing human needs and ecosystem function. *Frontiers in Ecology and the Environment, 2*(5), 249–57.

Eisen, L., Barker, C. M., Moore, C. G., Pape, W. J., Winters, A. M., & Cheronis, N. (2010). Irrigated agriculture is an important risk factor for West Nile virus disease in the hyperendemic Larimer–Boulder–Weld area of North Central Colorado. *Journal of Medical Entomology, 47*(5), 939–51.

Gibbs, H. K., Ruesch, A .S., Achard, F., Clayton, M. K., Holmgren, P., Ramankutty, N., & Foley, J. A. (2010). Tropical forests were the primary sources of new agricultural land in the 1980s and 1990s. *Proceedings of the National Academy of Sciences, 107*(38), 16732–7.

Goldberg, T. L., Gillespie, T., Rwego, I., Estoff, E., & Chapman, C. (2008). Forest fragmentation as cause of bacterial transmission among nonhuman primates, humans, and livestock, Uganda. *Emerging Infectious Diseases, 14*(9), 1375–82.

Goldberg, T., Paige, S., & Chapman, C. (2012). The Kibale EcoHealth Project: Exploring connections among human health, animal health, and landscape dynamics in Western Uganda. In A. A. Aguirre, R. S. Ostfeld, & P. Daszak (Eds.), *New directions in conservation medicine: Applied cases in ecological health* (pp. 452–65). New York, NY: Oxford University Press.

Gottdenker, N. L., Streicker, D. G., Faust, C. L., & Carroll, C. R. (2014). Anthropogenic land use change and infectious diseases: A review of the evidence. *EcoHealth, 11,* 619–32.

Hahn, M. B., Gurley, E. S., Epstein, J. H., Islam, M. S., Patz, J. A,. Daszak, P., & Luby, S. P. (2014). The role of landscape composition and configuration on Pteropus giganteus roosting ecology and Nipah virus spillover risk in Bangladesh. *American Journal of Tropical Medicine and Hygiene, 90*(2), 247–55.

Jones, K. E., Patel, N. G., Levy, M. A., Storeygard, A., Balk, D., Gittleman, J. L., & Daszak, P. (2008). Global trends in emerging infectious diseases. *Nature, 451*(7181), 990–3.

Karesh, W. B., & Noble, E. (2009). The bushmeat trade: Increased opportunities for transmission of zoonotic disease. *Mount Sinai Journal of Medicine, 76,* 429–34.

Karesh, W. B., Dobson, A., Lloyd-Smith, J. O., Lubroth, J., Dixon, M. A., Bennett, M., et al. (2012). Ecology of zoonoses: Natural and unnatural histories. *Lancet, 380*(9857), 1936–45.

Keiser, J., Maltese, M. F., Erlanger, T. E., Bos, R., Tanner, M., Singer, B. H., & Utzinger, J. (2005). Effect of irrigated rice agriculture on Japanese encephalitis, including challenges and opportunities for integrated vector management. *Acta Tropica, 95*(1), 40–57.

Laporte, N. T., Stabach, J. A., Grosch, R., Lin, T. S., & Goetz, S. J. (2007). Expansion of industrial logging in Central Africa. *Science, 316*(5830), 1451.

LeBreton, M., Pike, B., Saylors, K., Le Doux Diffo, J., Fair, J. N., Rimoin, A. W., et al. (2012). Bushmeat and infectious disease emergence. In A. A. Aguirre, R. S. Ostfeld, & P. Daszak (Eds.), *New directions in conservation medicine: Applied cases in ecological health* (pp. 164–78). New York, NY: Oxford University Press.

LoGiudice, K., Ostfeld, R. S., Schmidt, K. A., & Keesing, F. (2003). The ecology of infectious disease: Effects of host diversity and community composition on Lyme disease risk. *Proceedings of the National Academy of Sciences of the USA, 100*(2), 567–71.

Luby, S. P., Gurley, E. S., & Hossain, M. J. (2009). Transmission of human infection with Nipah virus. *Clinical Infectious Diseases, 49*(11), 1743–8.

Luby, S. P., Rahman, M., Hossain, M. J., Blum, L. S., Husain, M. M., Gurley, E., et al. (2006). Foodborne transmission of Nipah virus, Bangladesh. *Emerging Infectious Diseases, 12*(12), 1888–94.

Mather, T. N. (1993). The dynamics of spirochete transmission between ticks and vertebrates. In H. Ginsberg (ed.), *Ecology and environmental management of lyme Disease* (pp. 43–62). New Brunswick, NJ: Rutgers University Press.

McManus, D. P., Gray, D. J., Li, Y., Feng, Z., Williams, G. M., Stewart, D., et al. (2010). Schistosomiasis in the People's Republic of China: The era of the Three Gorges Dam. *Clinical Microbiology Reviews, 23*(2), 442–66.

Nasi, R., Taber, A., & Van Vliet, N. (2011). Empty forests, empty stomachs? Bushmeat and livelihoods in the Congo and Amazon Basins. *International Forestry Review, 13*(3), 355–68.

Nupp, T. E., & Swihart, R. K. (1996). Effect of forest patch area on population attributes of white-footed mice (Peromyscus leucopus) in fragmented landscapes. *Canadian Journal of Zoology, 74*, 467–72.

Ostfeld, R. S., & Keesing, F. (2000). Biodiversity and disease tisk: The case of lyme disease. *Conservation Biology, 14*(3), 722–8.

Ostfeld, R. S., Glass, G. E., & Keesing, F. (2005). Spatial epidemiology: An emerging (or re-emerging) discipline. *Trends in Ecology & Evolution, 20*(6), 328–36.

Ostfeld, R. S., Keesing, F., & Eviner, V. (2008). *Infectious disease ecology: Effects of ecosystems on disease and of disease on ecosystems.* Princeton, NJ: Princeton University Press.

Patz, J. A., Graczyk, T. K., Geller, N., & Vittor, A. Y. (2000). Effects of environmental change on emerging parasitic diseases. *International Journal for Parasitology*, *30*(12–13), 1395–405.

Patz, J. A., Olson, S. H., Uejio, C. K., & Gibbs, H. K. (2008). Disease emergence from global climate and land use change. *Medical Clinics of North America*, *92*(6), 1473–91.

Poulsen, J. R., Clark, C. J., Mavah, G., & Elkan, P. W. (2009). Bushmeat supply and consumption in a tropical logging concession in northern Congo. *Conservation Biology*, *23*(6), 1597–608.

Reiter, P. (2001). Climate change and mosquito-borne disease. *Environmental Health Perspectives*, *109*(Suppl 1), 141–61.

Salkeld, D. J., Padgett, K. A., & Jones, J. H. (2013). A meta-analysis suggesting that the relationship between biodiversity and risk of zoonotic pathogen transmission is idiosyncratic. *Ecology Letters*, *16*(5), 679–86.

Steinmann, P., Keiser, J., Bos, R., Tanner, M., & Utzinger, J. (2006). Schistosomiasis and water resources development: Systematic review, meta-analysis, and estimates of people at risk. *The Lancet Infectious Diseases*, *6*(7), 411–25.

Walsh, J., Molyneux, D., & Birley, M. (1993). Deforestation: Effects on vector-borne disease. *Parasitology*, *30*, 55–75.

Wilson, M., Cheke, R., Flasse, S., Grist, S., Osei-Ateweneboana, M., Tetteh-Kumah, A., et al. (2002). Deforestation and the spatio-temporal distribution of savannah and forest members of the Simulium damnosum complex in southern Ghana and south-western Togo. *Transactions of the Royal Society of Tropical Medicine and Hygiene*, *96*, 632–9.

Wolfe, N. D., Daszak, P., Kilpatrick, A. M., & Burke, D. S. (2005). Bushmeat hunting, deforestation, and prediction of zoonotic disease emergence. *Emerging Infectious Diseases*, *11*(12), 1822–7.

Young, H. S., Dirzo, R., Helgen, K. M., McCauley, D. J., Billeter, S. A., Kosoy, M. Y., et al. (2014). Declines in large wildlife increase landscape-level prevalence of rodent-borne disease in Africa. *Proceedings of the National Academy of Sciences*, *111*(19), 7036–41.

Zhu, H.-M., Xiang, S., Yang, K., Wu, X.-H., & Zhou, X.-N. (2008). Three Gorges Dam and its impact on the potential transmission of schistosomiasis in regions along the Yangtze River. *EcoHealth*, *23*(2), 137–48.

Chapter Six

Physical activity

The chronic health conditions endemic in many industrialized countries now account for the "leading causes of death globally, killing more people each year than all other causes combined" (WHO, 2011, p. 1). Regular physical activity can reduce the risk of chronic health conditions such as cardiovascular disease, type 2 diabetes, and selected forms of cancer, and it can also reduce the risk of osteoporosis as well as improve mental health and mood and increase longevity (CDC, 2011b). Considering all of the negative health consequences that can arise from a lack of physical activity, it has been estimated that physical *in*activity is now the fourth leading cause of death worldwide (Kohl et al., 2012). The level of regular physical activity associated with health benefits can be achieved in two ways. First, it can be achieved with 150 minutes/week of moderate intensity aerobic activity. Walking briskly is an example of moderate activity where the heart rate is raised without breathing hard. Second, regular physical activity can also be achieved performing more vigorous activities 75 minutes/week. Vigorous activities, such as running, elevate the heart rate to a degree that causes one to breathe hard.[1]

The ecosystem services presented thus far have been a product of the presence of GI in the local, regional, and global environment. What we consider now is how access to GI as an environmental amenity can facilitate the health behavior of physical activity. The ecological model of health recognizes the role of the environment in determining health behaviors, operating under the premise "that the environment largely controls or sets limits on the behavior that occurs in it," and "changing environmental variables results in the modification of behavior" (Green, Richard, & Potvin, 1996, p. 272). The environment that has received the most attention in decades of health promotion efforts aimed at encouraging physical activity has been the social environment. What has been lacking, until somewhat recently, has been a recognition of the fundamental role of the physical environment in determining behavior. The physical environment is an essential ingredient in addressing the pandemic of physical inactivity (Kohl et al., 2012; McKinnon, Reedy, Handy, & Rodgers, 2009).

There are a number of theories which support that GI, as an element of the physical environment, can be an important determinant in a holistic approach to health promotion aimed at increasing physical activity. Parks and public lands can act as *behavior settings* (Barker, 1963) for physical activity. "Behavior settings are the physical and social contexts in which behavior occurs" (Sallis, Bauman, & Pratt, 1998, p. 380). It is the perceived properties of a particular behavior setting that determine what possibilities it affords. "Affordances are the *functional* properties of an environmental feature for an individual" (Heft, 2010, p. 20). The perceived affordances of parks and other forms of GI determine their potential as behavior settings, or "arenas for action" (Heft, 2010, p. 28), for physical activity. If GI can be designed in such a way that it is perceived as affording the action of physical activity, then it can contribute to creating a supportive environment for physical activity (Sallis et al., 1998). A supportive environment includes attractive public spaces where physical activity, including actions as simple as walking, can be performed (Bedimo-Rung, Mowen, & Cohen, 2005; Coombes, Jones, & Hillsdon, 2010).

There has been nothing less than a small explosion in the literature over the past two decades applying these theories to better understand the influence of physical environment on physical activity behavior. This was spurred by the need to address the growing obesity epidemic and the concurrent recognition of the understudied influence of the physical environment in facilitating or deterring physical activity. This body of literature has largely focused on elements of the built environment such as land uses, block lengths, and road patterns. While our understanding of the relationship between the built environment and physical activity has grown, a systematic literature review on this topic concluded that more studies are greatly needed (Ferdinand, Sen, Rahurkar, Engler, & Menachemi, 2012). A subset of the built environment and physical activity literature has posited GI as a potential environmental support that can encourage the myriad health benefits gained through regular physical activity. Reviews of this literature have concluded that, overall, "the value of greenspaces as places to exercise is unquestionable"

(Croucher, Myers, & Bretherton, 2007, p. 26), but much more work needs to be done to understand the characteristics of greenspaces that encourage or deter its use. A small but growing body of evidence has begun to reveal some of these characteristics.

The general pathway by which parks influence use is presented in Figure 6.1. For a select few, the mere presence of a park in one's community may be enough to spur one to be active. For most, it is the characteristics of the park that will facilitate or deter its use for some form of physical activity. The characteristics of GI (e.g. size, aesthetic appeal) and the characteristics of the area surrounding GI are important in determining the type of activities that GI supports, from recreational walking to bicycle commuting.

The ability of GI to support physical activity is influenced foremost by access, which is most often measured by proximity. Once access is achieved, there are a number of other characteristics that influence its use. Using parks for walking is one such case where we see this play out. Overall use of parks for walking is associated with reduced distance to a park, and, after access is achieved, the characteristic of size appears to be more important than attractiveness (Giles–Corti et al., 2005). The size of the park trumps the features that make the space attractive

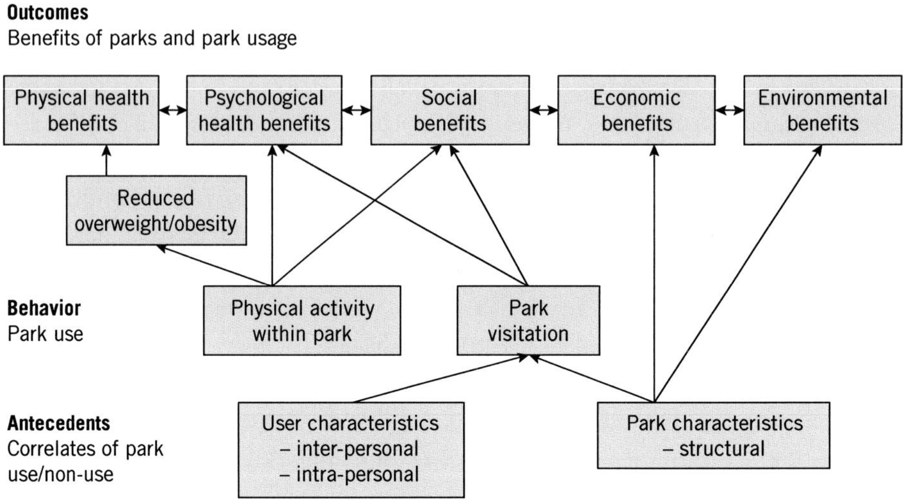

Outcomes
Benefits of parks and park usage

Figure 6.1: Determinants of park use and outcomes
Source: Adapted from Bedimo-Rung et al. (2005).

(e.g. shade, water features, quietness, lighting, birdlife). This has been supported in a study that revealed that open space near home is associated with all levels of recreational walking, but large, high-quality open spaces are predictive of achieving the levels of physical activity needed to promote health benefits (Sugiyama, Francis, Middleton, Owen, & Giles-Corti, 2010). Noting quality also matters suggests that attractiveness should not be overlooked. In fact, it is encouraging to note that observable biodiversity is one such attractive feature that people seek when using a park for activity (Bird, 2004). Biodiversity not only encourages the health benefits associated with the use of GI but also improves the ability of GI to deliver other ecosystem services. Attractiveness measured by its perceived naturalness and other aesthetic qualities is critical, but size likely needs to be considered first. Larger greenspaces can support a wider variety of activities that can lead to the levels of physical activity needed to achieve health benefits.

Recreational facilities that are not green may also support physical activity, but these recreational facilities do not appear to have the same impact as GI in encouraging physical activity. A useful summary of the parks and physical activity literature by the type of setting allows us to isolate green settings (trails, parks, open space, and water fronts) from other recreational facilities (Kaczynski & Henderson, 2007). Settings with natural qualities were stronger predictors of physical activity than other recreational facilities. This is likely because green recreational facilities such as parks allow persons of various ages, abilities, and preferences to use the GI for a variety of activities, unlike some sports fields that may be used for a variety of purposes but are designed for only one. There is a preference for greenspaces with more natural qualities as opposed to sports fields (Backlund, Stewart, McDonald, & Miller, 2004). This speaks to the biophilic draw of the natural qualities of less adulterated GI.

Size is critical, but restoring or conserving large patches of GI in urban environments where space is limited may simply not be possible. Space constraints can be overcome by creating a system of GI that connects patches of various sizes to one another with corridors. Here, size is increased not by expanding any given greenspace, but by connecting and expanding the GI system. The GI system serves the dual function of creating the landscape structure necessary to support ecosystem functioning, and it creates a more accessible system for bike and pedestrian activity. One may not live or work near a large patch of GI, but greenway corridors branching out through the built environment and connecting patches brings a part of the GI system closer to more people and provides a means to access all other parts of the system.

A greenway is defined as a "linear open space established along either a natural corridor, such as a riverfront, stream valley, or ridgeline, or overland along a railroad right-of-way converted to recreational use, a canal, a scenic road, or other route" or an "open space connector linking parks, nature reserves, cultural features, or historic sites with each other and with populated areas" (Figure 6.2) (Little, 1989, p. 1). An increasing number of greenway projects are being adopted internationally (Fabos & Ahern, 1996; Lindsey, 1999) with the potential for significant use by persons performing multiple forms of non-motorized locomotion (Lindsey & Nguyen, 2004). A greenway can be viewed as a GI corridor that ideally connects parks or other components of a GI system. It is a type of trail, but a trail can only be considered a greenway if it creates a natural corridor by conserving adjacent land. Greenways do not need to include multi-use trails in their design, but they often do. In doing so, they maximize the affordances of the greenway to support physical activity.

Figure 6.2: Neuse River Greenway, Raleigh, North Carolina
Source: James Willamor.

Greenways are unique in that it is not only access to the greenway that may influence its use but also the accessibility of the places along the greenway route. Greenways are more likely to be used regularly if they connect people to the places they live and want to go. Connecting commercial and residential areas influences whether activities, such as walking, are seen as an option (Handy, 1996). A number of studies have measured the pedestrian accessibility of locations on street networks (Aultman-Hall, Roorda, & Baetz, 1997; Talen, 2003), but this network approach has been seldom applied to GI even though GI has the same potential to intersect a variety of land uses and create connections between potential origins and destinations. It is true that connecting origins and destinations may not be as critical to recreational activity achieved on a greenway, but, if we are trying to fold physical activity into people's daily lives, active utilitarian travel must be encouraged.

The heightened accessibility of places facilitated by a GI system that connects potential origins and destinations is critical to encourage physical activity achieved through utilitarian walking and biking. One such form of utilitarian physical activity is commuting to work. Having local multi-use trails has been shown to be associated with non-motorized commuting activity (Nelson & Allen, 1997), and the accessibility that "localness" creates appears to be important for commuting. In a study comparing the intensity of use on local, regional, and state greenways, the highest percentage of commuters (12 percent) were found on local trails (Gobster, 1995). What can be inferred from this is that people on local trails had the right combination of factors working together to make commuting on a greenway feasible—their workplace was within bike commuting distance, and a local greenway, at least approximately, connected their home and work locations. Since local trails are more often found in areas that have a higher degree of land-use mixture (compared to regional trails that extend into the countryside), it is possible that the increased level of opportunities created by a greater land-use mixture is what was influencing the higher percentage of use for commuting. When trails are local, they are also used frequently. Fifty nine percent of local greenway users claimed they use local trails "virtually every day" or "virtually every week" (Gobster, 1995). Cyclists traveling primarily on separated paths on their journey to work are also willing to make significantly longer trips (Shafizadeh & Niemeier, 1997). In other words, when paths are available, people are willing to travel a little further, achieving more physical activity in the process. Commuting is just one form of utilitarian travel, and it is a tricky one considering the complicated spatial relationship between where one lives and works may not be able to be reconciled by a greenway. Still, a connected GI system increases the likelihood of connecting home and work and making an active commute possible. Furthermore, a connected GI system also increases the likelihood that the system will intersect a greater variety of land uses and therefore connect more potential origins and destinations. This is essential to encourage other forms of active utilitarian trips such as daily shopping or a trip to the post office (Figure 6.3).

Not overlooking the fact that parks are a type of land use, and strengthening the case for connecting greenways to parks, large parks have been found to be hubs of activity on urban greenway systems (Coutts, 2009). More parkland acreage has also been found to be significantly correlated with increased utilitarian walking and bicycling (Zlot & Schmid, 2005). People get physical activity on utilitarian trips in areas with more parkland possibly because this type of land use makes an active utilitarian trip more pleasant. Parks could also be an attractive stopping off point or themselves a forum for activity in environments with greater land-use mixture. This was the case for older adults in Bogota, Columbia, where land-use

Figure 6.3: An urban greenway in Minneapolis connects a variety of land uses including an adjacent community garden

Source: Payton Chung.

mixture and higher park density were associated with active park use (Parra, Gomez, Fleischer, & David Pinzon, 2010). In addition, speaking not to land use but the demographic makeup of the various neighborhoods an urban greenway may intersect, greenway users appear to be uninhibited by the racial characteristics of the neighborhoods they pass through (Coutts & Miles, 2011). Greenways not only create connections between potential origins and destinations but also between neighborhoods and people.

Density of development also appears to be an essential ingredient for getting more people to use GI for physical activity. Connecting a variety of land uses with a GI system is good, but if land uses are spread over a large geographic area, the greater distances between potential origins and destinations may deter some people from choosing active travel. Denser development that incorporates natural recreation areas would prevent the extent of automobile-dependence and create opportunities to be physically active (Frumkin, 2002). Denser development not only brings origins and destinations closer to one another, but it also often creates higher population densities. The interaction between higher levels of land-use mixture and population density in areas surrounding greenways has been found to positively influence the use of greenways for physical activity (Coutts, 2008). Greenways are a form of GI that can intersect areas of high residential density and land-use mixture without consuming the space that would disperse development and reduce density.

The body of evidence linking GI and physical activity is laden with associations. Therefore, there is little that addresses the question of whether GI actually causes more physical activity in people who are already active or encourages inactive people to become active. For example, it has been found that trail users who use trails for both recreational and utilitarian purposes are generally more physically active than non-trail users (Librett, Yore, & Schmid, 2006). Building off this association, the question is then "How do we get more people to become trail users?" It does not appear to be by building more trails. A prospective study done by Evenson, Herring, and Huston (2005) found no relationship between the creation of a trail on a converted rail line and increased levels of physical activity. Simply building more trails may not be the answer, but we need to do a much better job of understanding the characteristics and contexts of GI in order to make it a better prescription for getting people active. This is reflected in calls that stress the need to better understand the mediating and moderating mechanisms in GI and health research (Lachowycz & Jones, 2012) and potential research pitfalls (Oakes, Mâsse, & Messer, 2009).

Some of work from the Netherlands that has examined GI for its role in mediating the relationship between physical activity and health found that the stress reduction and social cohesion properties of GI were more important to health than physical activity performed in greenspace (deVries, van Dillen, Groenewegen, & Spreeuwenberg, 2013). Mental health (specifically stress), physical activity, and social capital or cohesion are the "big three" that have garnered the majority of the attention in research examining the relationship between health and the built and natural environments. While there are certainly relationships between these three determinants of health, they are most often examined separately. All have been found to be important separately and improved by GI (with caveats, of course). What de Vries et al. (2013) uncovered was that stress reduction and social cohesion may be the greatest health benefits of GI, not physical activity. This confirms their previous suspicion that "stress reduction and the facilitation of social cohesion are likely to be more important than … stimulating physical activity" (de Vries, 2010, p. 87). So, the generally positive results of research showing a relationship between GI and physical activity may not translate into improved health, and physical activity may not be as important as the other benefits of GI. A critical factor that must be taken into account with this work from the Netherlands, and all other GI and health research, is the context of where the studies were conducted.

Differences in attitudes towards physical activity and the design of the built environment in countries where studies have been conducted temper our ability to draw any definitive universal conclusions about the ability of GI to promote physical activity. For example, while there is evidence to support that increased access to greenspace positively influences overall levels of physical activity in the USA (Cohen et al., 2007) and walking in Australia (Giles-Corti et al., 2005), there are studies from the UK and the Netherlands that have found no relationship (Hillsdon, Panter, Foster, & Jones, 2006; Maas, Verheij, Spreeuwenberg, & Groenewegen, 2008). This may be due to the fact that the environment in countries such as the Netherlands, with much more infrastructure devoted to pedestrians and cyclists, is generally much more conducive to physical activity. Unlike some countries where GI may be the only safe place to perform physical activity, in the Netherlands GI is just one of the many forums available for activity. One of the reasons that we love the Netherlands is that there are many civic environment supports for daily recreational and utilitarian physical activity. It is in places such as this where GI is possibly not as critical as a forum for physical activity, but where other co-benefits can be the focus.

The mixed, but positively leaning, results on the ability of GI to promote physical activity mirrors the assortment of benefits particular populations believe that GI affords them. In some communities, the perceived benefits of trails include opportunities to exercise (Shafer, Lee, & Turner, 2000). In others, it is "more often the 'de-stressing' capacity of greenspaces, their role as an 'escape' from the dirt, stress, noise, and visual hardness of urban settings and their restorative capacity [that] are the most valued factors" (Croucher et al., 2007, p. 20). The benefits of parks to overall health and social well-being vary by race and ethnicity in US cities (Ho, Sasidharan, & Elmendorf, 2005) and also in Hong Kong where use and attitudes towards urban GI vary according to the "social profiles" of residential communities (Lo & Jim, 2010). Policies that create environmental supports for physical activity, such as active travel, appear promising, but much remains unclear about how GI can complement other physical supports for promoting active travel in particular populations (Fraser & Lock, 2011).

To accommodate a variety of attitudes and preferences, there may be a need to be more creative in considering the ways that people are physically active in their interactions with GI. There are many ways that GI can promote physical activities beyond walking and biking. Green infrastructure conservation is one such activity that benefits both GI and the individuals doing the manual labor of environmental conservation. This has been embraced on the landscape scale in the Green Gyms movement (The Conservation Volunteers, 2014). Green Gyms involve persons getting out doors and involved in the social and physical aspects of environmental conservation. Unlike other environmental conservation efforts focused on protecting the environment for the environment's sake, the Green Gym movement treats the conservation activities of clearing land, path building, planting, and digging as physical activities. Similar to lifting weights in gym, this includes warm up and cool down periods. There are the social, physical, and mental health benefits of the activities themselves, and there are also the many health benefits derived from the restoration of GI and the ecosystems services it supports.

Overweight and obesity

One of the major drivers to create environments conducive to physical activity is the recognized need to stem the epidemic of obesity most pronounced in the USA but becoming a global issue. Over one-third of adults and nearly 20 percent of children in the USA are obese (Ogden, Carroll, Kit, & Flegal, 2014). Worldwide, obesity has nearly doubled since 1980 (WHO, 2015). Most of the recent research aimed at understanding and altering environments to support greater levels of physical activity have been motivated by the assumption that more physical activity will prevent many of the diseases associated with overweight and obesity. Some of the environment and physical activity research has specifically examined GI for its potential to support physical activity and the expenditure of calories necessary to reduce overweight and obesity. While getting active is more beneficial to health than reducing obesity (Ekelund et al., 2015), obesity is still an increasingly prevalent precursor to many preventable health issues.

Even modest amounts of physical activity, such as the levels recommended by the Centers for Disease Control and Prevention (CDC), can prevent increases in visceral fat (fat in the abdominal cavity), a well-documented major risk factor for diseases such as type 2 diabetes, cardiovascular disease, metabolic disturbances, and some cancers (Slentz et al., 2005; Harvard Medical School, 2007). Slentz et al. (2005) found that regular physical activity prevented significant increases in visceral fat, and getting a bit more than the CDC recommended dose of exercise significantly decreased visceral, subcutaneous (waist fat, the stuff you can pinch), and total abdominal fat (visceral and subcutaneous) *without a change in diet*. Correct, without a change in diet. This was supported in a study examining the correlates of the spike of obesity in the USA (Ladabaum, Mannalithara, Myer, & Singh, 2014). Caloric intake did not change greatly between 1988 and 2010, but physical activity dropped precipitously with the concurrent rise in obesity. On the other hand, diet has also been shown to supersede activity (Carroll, 2015), but the litany of other health benefits of regular physical activity should caution against focusing on diet alone.

Even if physical activity can overcome diet, this in no way marginalizes the complementary and expanding research on food environments (Figure 6.4). Proper nutrition and its associated health effects depend on the availability of certain types of food and not just the amount of calories consumed. Also, food environments are an essential ingredient to reducing overweight and obesity given that reaching the levels of physical activity necessary to overcome a poor

Figure 6.4: All you can eat, Amsterdam
Source: Rex Roof.

diet will not be the answer for everyone (e.g. persons with physical limitations). The environmental determinants of food choices and physical activity behavior need to be considered together in order to fully understand obesegenic environments (Lake & Townshend, 2006). The literature has been much more focused on the ability of the environment to increase caloric expenditure and less so on reducing the calories consumed. Therefore, we know very little about how GI can contribute to creating a healthy food environment (e.g. community gardens) and only slightly more about its role in alleviating obesity.

Although there is some evidence to suggest that increased access to GI lowers the likelihood of obesity (Nielsen & Hansen, 2007), in a review of the literature on greenspace and obesity Lachowycz and Jones (2011) found that the results were generally positive but too inconsistent to conclude that greenspace in one's environment leads to lower rates of overweight and obesity. Much like our nascent understanding of the built and natural environment variables that support physical activity, the array of measures employed to examine the physical environment correlates of obesity have made any definitive conclusions elusive (Booth, Pinkston, & Poston, 2005). Since physical activity plays such a key role in reducing obesity, our ability to connect GI to reduced obesity would be improved by examining GI, physical activity, and obesity together to elucidate the pathway by which having accessible GI leads to physical activity that then leads to reduced obesity. When examining GI, physical activity, and obesity together, it was shown that more neighborhood greenery was associated with both more physical activity and reduced levels of self-reported overweight and obesity (Ellaway, Macintyre, & Bonnefoy, 2005). It can be assumed that the physical activity brought about by more greenery was having the mediating effect on reducing overweight and obesity, but this cannot be concluded definitely.

Looking at a characteristic of the built environment, GI, and obesity together was an examination of how residential density and the amount of greenspace influences children's body mass index (BMI) (Bell, Wilson, & Liu, 2008). Population and residential density have been used as proxies to represent built environment types that support physical activity; the belief being that areas with greater density are more likely to include the physical features (e.g. reduced proximities, land-use mixture, more connected street networks) that support walking and biking. This was not found to be the case for adults in a study of residential density and the levels of walking for physical activity necessary to provide health benefits and reduce obesity (Forsyth, Oakes, Schmitz, & Hearst, 2007).[2] Likewise, in the Bell et al. (2008) study of children, residential density alone was shown to have no effect on children's BMI, *but* the greenness of neighborhoods was associated with lower BMI regardless of residential density. Furthermore, more greenness reduced the odds of children increasing their BMI over a two-year period. As the authors note, the mechanism that may explain this is that children take advantage of more types of GI (e.g. yards, parks, vacant lots) than adults. Getting kids more active and reducing childhood overweight and obesity is critical not only to childhood health, but also because it has consequences for physical activity behavior later in life. More active children leads to more active adults (Telama et al., 2005).

Summary

The slowly growing body of GI and physical activity research has resulted in mixed, but positively leaning, results demonstrating that GI is supportive of the health behavior of physical activity. While an increasingly sophisticated array of measures is being used to advance the science (Sallis, 2009), there is sufficient evidence to support policies that conserve and expand GI for the purpose of promoting physical activity. This is especially true considering the myriad co-benefits of doing so. The ability of GI to support physical activity may not be as important as its other ecosystem services in places where the built environment supports physical activity, but what does appear universal is that the characteristics of GI, and of areas surrounding GI, influence its use for active recreation and transportation. Accessibility and size are two essential characteristics that can simultaneously be increased by creating an interconnected system of GI. Though the complexity of environmental interventions such as conserving and restoring GI may seem "daunting and perhaps unachievable" as a public health intervention, it is essential to an ecological approach aimed at promoting health behavior (Stokols, 1992, p. 6).

Notes

1. These aerobic activities once constituted the whole of what was considered regular physical activity, but the Centers for Disease Control and Prevention (CDC) now also recommends two hours/week of muscle strengthening exercise (CDC, 2011a).
2. The cross-sectional nature of this study, and others, always introduces the possibility of residential self-selection and the possibility of a type 1, or, in this case, a type 2 error.

References

Aultman-Hall, L., Roorda, M., & Baetz, B. W. (1997). Using GIS for evaluation of neighborhood pedestrian accessibility. *Journal of Urban Planning and Development, 123*(1), 211–25.

Backlund, E. A., Stewart, W. P., McDonald, C., & Miller, C. (2004). Public evaluation of open space in Illinois: Citizen support for natural area acquisition. *Environmental Management, 34*(5), 634–41.

Barker, R. G. (1963). On the nature of the environment. *Journal of Social Issues, 19*(4), 17–38.

Bedimo-Rung, A. L., Mowen, A. J., & Cohen, D. A. (2005). The significance of parks to physical activity and public health: A conceptual model. *American Journal of Preventive Medicine, 28*(2 Suppl 2), 159–68.

Bell, J. F., Wilson, J. S., & Liu, G. C. (2008). Neighborhood greenness and 2-year changes in body mass index of children and youth. *American Journal of Preventive Medicine, 35*(6), 547–53.

Bird, W. (2004). *Natural fit: Can green space and biodiversity increase levels of physical activity?* Sandy, UK: Royal Society for the Protection of Birds.

Booth, K. M., Pinkston, M. M., & Poston, W. S. C. (2005). Obesity and the built environment. *Journal of the American Dietetic Association, 105*(5 Suppl 1), S110–17.

Carroll, A. E. (2015). To lose weight, eating less is far more important than exercising more. *New York Times,* June 15.

CDC. (2011a). How much physical activity do adults need? Retrieved November 19, 2013, from www.cdc.gov/physicalactivity/everyone/guidelines/adults.html

CDC. (2011b). Physical activity for everyone: The importance of physical activity. Retrieved from www.cdc.gov/physicalactivity/everyone/health/

Cohen, D., McKenzie, T., Sehgal, A., Williamson, S., Golinelli, D., & Lurie, N. (2007). Contribution of public parks to physical activity. *American Journal of Public Health, 97*(3), 509–14.

Coombes, E., Jones, A. P., & Hillsdon, M. (2010). The relationship of physical activity and overweight to objectively measured green space accessibility and use. *Social Science & Medicine, 70*(6), 816–22.

Coutts, C. (2008). Greenway accessibility and physical-activity behavior. *Environment and Planning B: Planning and Design, 35*(3), 552–63.

Coutts, C. (2009). Multiple case studies of the influence of land-use type on the distribution of uses along urban river greenways. *Journal of Urban Planning and Development, 135*(1), 31–8.

Coutts, C., & Miles, R. (2011). Greenways as green magnets: The relationship between the race of greenway users and race in proximal neighborhoods. *Journal of Leisure Research, 43*(3), 317–33.

Croucher, K., Myers, L., & Bretherton, J. (2007). *The links between greenspace and health: A critical literature review.* Stirling, UK: Greenspace Scotland.

de Vries, S. (2010). Nearby nature and human health: Looking at mechanisms and their implications. In C. Ward Thompson, P. Aspinall, & S. Bell (Eds.), *Innovative approaches to researching landscape and health: Open space: People space 2* (pp. 77–96). Abingdon: Routledge.

de Vries, S., van Dillen, S. M. E., Groenewegen, P. P., & Spreeuwenberg, P. (2013). Streetscape greenery and health: Stress, social cohesion and physical activity as mediators. *Social Science & Medicine, 94,* 26–33.

Ekelund, U., Ward, H. A., Norat, T., Luan, J., May, A. M., Weiderpass, E., et al. (2015). Physical activity and all-cause mortality across levels of overall and abdominal adiposity in European men and women: The European Prospective Investigation into Cancer and Nutrition study (EPIC). *American Journal of Clinical Nutrition, 101*(C), 613–21.

Ellaway, A., Macintyre, S., & Bonnefoy, X. (2005). Graffiti, greenery, and obesity in adults: Secondary analysis of European cross sectional survey. *BMJ (Clinical Research Ed.), 331*(7517), 611–12.

Evenson, K. R., Herring, A. H., & Huston, S. L. (2005). Evaluating change in physical activity with the building of a multi-use trail. *American Journal of Preventive Medicine, 28*(2 Suppl 2), 177–85.

Fabos, J., & Ahern, J. (Eds.). (1996). *Greenways: The beginning of an international movement.* Amsterdam: Elsevier Science.

Ferdinand, A. O., Sen, B., Rahurkar, S., Engler, S., & Menachemi, N. (2012). The relationship between built environments and physical activity: A systematic review. *American Journal of Public Health, 102*(10), e7–e13.

Forsyth, A., Oakes, J. M., Schmitz, K. H., & Hearst, M. (2007). Does residential density increase walking and other physical activity? *Urban Studies, 44*(4), 679–97.

Fraser, S. D. S., & Lock, K. (2011). Cycling for transport and public health: A systematic review of the effect of the environment on cycling. *European Journal of Public Health, 21*(6), 738–43.

Frumkin, H. (2002). Urban sprawl and public health. *Public Health Reports, 117*(3), 201.

Giles-Corti, B., Broomhall, M. H., Knuiman, M., Collins, C., Douglas, K., Ng, K., et al. (2005). Increasing walking: How important is distance to, attractiveness, and size of public open space? *American Journal of Preventive Medicine, 28*(2 Suppl 2), 169–76.

Gobster, P. (1995). Perception and use of a metropolitan greenway system for recreation. *Landscape Planning, 33*(1), 401–13.

Green, L. W., Richard, L., & Potvin, L. (1996). Ecological foundations of health promotion. *American Journal of Health Promotion, 10*(4), 270–81.

Handy, S. L. (1996). Methodologies for exploring the link between urban form and travel behavior. *Transportation Research. Part D, Transport and Environment, 1*(2), 151–65.

Harvard Medical School. (2007). Abdominal fat and what to do about it. Retrieved May 27, 2015, from www.health.harvard.edu/fhg/updates/Abdominal-fat-and-what-to-do-about-it.shtml

Heft, H. (2010). Affordances and the perception of landscape: An inquiry into environmental perception and aesthetics. In C. Ward-Thompson, P. Aspinall, & S. Bell (Eds.), *Innovative approaches to researching landscape and health: Open space: People space 2* (pp. 9–32). Abingdon: Routledge.

Hillsdon, M., Panter, J., Foster, C., & Jones, A. (2006). The relationship between access and quality of urban green space with population physical activity. *Public Health, 120*(12), 1127–32.

Ho, C., Sasidharan, V., & Elmendorf, W. (2005). Gender and ethnic variations in urban park preferences, visitation, and perceived benefits. *Journal of Leisure Research, 37*(3), 281–306.

Kaczynski, A. T., & Henderson, K. A. (2007). Environmental correlates of physical activity: A review of evidence about parks and recreation. *Leisure Sciences, 29*(4), 315–54.

Kohl, H. W., Craig, C. L., Lambert, E. V., Inoue, S., Alkandari, J. R., Leetongin, G., & Kahlmeier, S. (2012). The pandemic of physical inactivity: Global action for public health. *The Lancet, 380*(9838), 294–305.

Lachowycz, K., & Jones, A. P. (2011). Greenspace and obesity: A systematic review of the evidence. *Obesity Reviews, 12*(5), e183–9.

Lachowycz, K., & Jones, A. P. (2012). Towards a better understanding of the relationship between greenspace and health: Development of a theoretical framework. *Landscape and Urban Planning,* 8–15.

Ladabaum, U., Mannalithara, A., Myer, P. A., & Singh, G. (2014). Obesity, abdominal obesity, physical activity, and caloric intake in U.S. adults: 1988–2010. *American Journal of Medicine, 127*(8), 717–27.

Lake, A., & Townshend, T. (2006). Obesogenic environments: Exploring the built and food environments. *Journal of the Royal Society for the Promotion of Health, 126*(6), 262–7.

Librett, J. J., Yore, M. M., & Schmid, T. L. (2006). Characteristics of physical activity levels among trail users in a U.S. national sample. *American Journal of Preventive Medicine, 31*(5), 399–405.

Lindsey, G. (1999). Use of urban greenways: Insights from Indianapolis. *Landscape and Urban Planning, 45*(2–3), 145.

Lindsey, G., & Nguyen, D. B. L. (2004). Use of greenway trails in Indiana. *Journal of Urban Planning and Development, 130*(4), 213–17.

Little, C. E. (1989). *Greenways for America*. Baltimore, MD: Johns Hopkins University Press.

Lo, A.Y. H., & Jim, C.Y. (2010). Differential community effects on perception and use of urban greenspaces. *Cities, 27*(6), 430–42.

Maas, J., Verheij, R. A., Spreeuwenberg, P., & Groenewegen, P. P. (2008). Physical activity as a possible mechanism behind the relationship between green space and health: A multilevel analysis. *BMC Public Health, 8,* 206.

McKinnon, R. A., Reedy, J., Handy, S. L., & Rodgers, A. B. (2009). Measuring the food and physical activity environments: Shaping the research agenda. *American Journal of Preventive Medicine, 36*(4 Suppl), S81–5.

Nelson, A. C., & Allen, D. (1997). If you build them, commuters will use them: Association between bicycle facilities and bicycle commuting. *Transportation Research Record, 1578,* 79–83.

Nielsen, T. S., & Hansen, K. B. (2007). Do green areas affect health? Results from a Danish survey on the use of green areas and health indicators. *Health & Place, 13*(4), 839–50.

Oakes, J. M., Mâsse, L. C., & Messer, L. C. (2009). Work group III: Methodologic issues in research on the food and physical activity environments: Addressing data complexity. *American Journal of Preventive Medicine, 36*(4 Suppl), S177–81.

Ogden, C. L., Carroll, M. D., Kit, B. K., & Flegal, K. M. (2014). Prevalence of childhood and adult obesity in the United States, 2011–2012. *Journal of the American Medical Association, 311*(8), 806–14.

Parra, D. C., Gomez, L. F., Fleischer, N. L., & David Pinzon, J. (2010). Built environment characteristics and perceived active park use among older adults: Results from a multilevel study in Bogotá. *Health & Place, 16*(6), 1174–81.

Sallis, J. F. (2009). Measuring physical activity environments: A brief history. *American Journal of Preventive Medicine, 36*(4 Suppl), S86–92.

Sallis, J. F., Bauman, A. E., & Pratt, M. (1998). Environmental and policy interventions to promote physical activity. *American Journal of Preventive Medicine, 15*(4), 379–97.

Shafer, C. S., Lee, B. K., & Turner, S. (2000). A tale of three greenway trails: User perceptions related to quality of life. *Landscape and Urban Planning, 49*(3–4), 163–78.

Shafizadeh, K., & Niemeier, D. (1997). Bicycle journey-to-work: Travel behavior characteristics and spatial attributes. *Transportation Research Record, 1578,* 84–90.

Slentz, C. A., Aiken, L., Houmard, J. A., Bales, C. W., Johnson, J. L., Tanner, C. J., et al. (2005). Inactivity, exercise, and visceral fat. STRRIDE: A randomized, controlled study of exercise intensity and amount. *Journal of Applied Physiology, 99*(4), 1613–18.

Stokols, D. (1992). Establishing and maintaining healthy environments: Toward a social ecology of health promotion. *American Psychologist, 47*(1), 6–22.

Sugiyama, T., Francis, J., Middleton, N. J., Owen, N., & Giles-Corti, B. (2010). Associations between recreational walking and attractiveness, size, and proximity of neighborhood open spaces. *American Journal of Public Health, 100*(9), 1752–7.

Talen, E. (2003). Neighborhoods as service providers: A methodology for evaluating pedestrian access. *Environment and Planning B: Planning and Design, 30*(2), 181–200.

Telama, R., Yang, X., Viikari, J., Välimäki, I., Wanne, O., & Raitakari, O. (2005). Physical activity from childhood to adulthood: A 21-year tracking study. *American Journal of Preventive Medicine, 28*(3), 267–73.

The Conservation Volunteers. (2014). Green gym. Retrieved May 27, 2015, from www.tcv.org.uk/greengym

WHO. (2011). *Global status report on noncommunicable diseases 2010.* Geneva: WHO.

WHO. (2015). Obesity and overweight. Retrieved January 20, 2015, from www.who.int/mediacentre/factsheets/fs311/en/

Zlot, A., & Schmid, T. (2005). Relationships among community characteristics and walking and bicycling for transportation or recreation. *American Journal of Health Promotion, 19*(4), 314–17.

Chapter Seven

Mental health

Up to this point, GI has been discussed for its role in supporting physical health, but fully achieving health requires striving for complete physical *and* mental health, mental health being "a state in which a person is most fulfilled, can make sense of their surroundings, feel in control, can cope with every day demands and has purpose in life" (Bird, 2007, p. 27). The two are inseparable, and physical ailments often manifest when mental health is compromised. There are a number of ways that exposure to and affiliation with GI have been shown to support mental health. These include nature's ability to reduce stress, create positive affective states, and improve cognitive functioning.

What most often comes to the fore as the empirical basis for why GI is good for mental health are the many decades of environmental psychology and environment–behavior research examining the mentally restorative potential of exposure to the natural environment and elements of nature. The innate human preference for the natural environment and processes endows the natural environment with the unique ability to restore and renew "diminished functional resources and capabilities" (Hartig & Staats, 2003, p. 103), not only permitting mental restoration but promoting it so that daily physical, psychological, and social demands can be met (Hartig, 2007). Nature is thought to be restorative due to millions of years of human evolution in environments dominated by nature having hard-wired us with a predisposition to respond positively to natural settings that can support survival (Figure 7.1) (Appleton, 1975; Kaplan & Kaplan, 1989; Orians, 1986; Ulrich, 1993). It is also important to note that there is great degree of nuance to this work, forcing one to take a cautious approach to assuming a solely

Figure 7.1: African savanna
Source: Gossipguy.

evolutionary or genetic basis, unchanged over eons, as being solely responsible for humans responding positively to nature and the natural environment. There is also a sociocultural component working in concert with the genetic component (Hartig, 1993). Applying what we now know of the powerful influence of epigenetics, or the influence of environment on the genetic expression of many conditions, the sociocultural environment likely influences the degree of psychological and restorative benefits that we may be genetically predisposed to receive. A genetic basis is likely since a lack of nature can have a negative effect on mental health even in individuals who do not express any cultural appreciation for plants and nature (Grinde & Patil, 2009), but even a genetic predisposition has likely changed over time, and its level of expression moderated by sociocultural factors.

The two dominant pathways by which nature is theorized as providing mental restoration are recovery from stress and recovery from attention fatigue.[1] Research examining the stress and attention antecedents and the benefits of recovery from each has largely been guided by psychoevolutionary (or psychophysiological) stress recovery theory (Ulrich, 1983) and attention restoration theory (ART) (R. Kaplan & Kaplan, 1989; S. Kaplan, 1995), respectively. Most of the restoration research has been guided by the psychophysiological stress recovery theory (Hartig &

Staats, 2003, p. 103)—leading to restoration and stress recovery to be incorrectly used interchangeably at times—but restoration is a broad concept "not limited to stress recovery situations or to recovery from excessive physiological arousal" (Ulrich, 1993). Stress recovery is not entirely restoration, but it is a large part of it. Restoration also encompasses recovery from attention fatigue which can result in a "lowered ability to concentrate and solve problems, heightened irritability, and a greater proneness to mistakes or accidents" (Herzog, Black, Fountaine, & Knotts, 1997, p. 165). Restoration theory does not invoke the biophilia hypothesis, but it is in line with biophilic tendency of nature to provide deep-seeded psychological needs (Grinde & Patil, 2009).

Restoration has often been examined through either the stress or attention lens, but these two concepts are not necessarily at odds and may even complement one another (Kaplan, 1995). They "occur alone in some circumstances, but in other circumstances they may have some form of reciprocal relationship or otherwise coincide" (Hartig, Evans, Jamner, Davis, & Gärling, 2003, p. 110). Directed attention fatigue has been posited as an aftereffect of stress (Cohen & Spacapan, 1978), and attention fatigue might make a person more vulnerable to stress (Kaplan, 1995) and its proven negative physical and mental health outcomes. Although the physiological connection between restoration from stress and attention has not been fully developed, there has been a case made for a possible relationship between parasympathetic nervous system responses to stress leading to reduced attention (Health Council of the Netherlands and Dutch Advisory Council for Research on Spatial Planning, 2004). The continued development of theories such as this will aid in our understanding of how stress and attention interact, but what we are concerned with here is how nature's ability to support stress and attention recovery translates into improved mental health. Our knowledge of the influence of stress on health is highly developed and continually expanding. The health benefits of recovery from attention fatigue much less so.

Although attention restoration theory (ART) has often been put forth as the basis of many studies exploring the mental health benefits of exposure to nature, there is seldom an explanation as to how nature's ability to restore directed attention produces health benefits. The heightened cognition brought about by having attention restored is certainly an attribute that most would find desirable, but it is unclear if this, in itself, makes us healthier. Nonetheless, ART is presented here for two reasons. First, because of the relationship between stress and attention. Attention fatigue can form in the absence of stress and this fatigue can then affect one's ability to cope with stressful situations (Kaplan & Kaplan, 1989). Second, it is also useful in helping to explain at least two behavioral attributes with health consequences that can be linked to cognition: directed attention (concentration) and inhibition. An example of one such behavior linked to an inability to concentrate is Attention Deficit Hyperactivity Disorder. The other attribute, inhibition, is important to our understanding of health behaviors because restoration from attention fatigue is more likely to place an inhibitory capacity under voluntary control, and thus is "an indispensable mechanism for behaving appropriately" (Kaplan, 1995, p. 171); "appropriately" in this case being in a way that promotes health.

We proceed here by first focusing on a handful of studies that explore the mental health benefits of exposure to nature that do not directly test whether these benefits are attributable to either stress recovery or attention restoration. This is followed by a focus on the relationship between stress and exposure to the natural environment. Stress has been shown to be an important mediator in the relationship between exposure to greenspace and health (de Vries, van Dillen, Groenewegen, & Spreeuwenberg, 2013), but, as we will see, there is also evidence to support the positive effect exposure to nature has on positive affect or emotions and improved attention. I therefore present how exposure to nature can result in positive affect and improved cognition with the basis of cognition being ART. If we are to continue to stay true to the accepted definition of health, improved affect certainly has its role to play in improving well-being, and recovery from attention fatigue undoubtedly has benefits in that directed attention supports our cognitive ability to achieve fulfillment and cope with everyday demands. Taking stress, affect, and attention together, there is sufficient evidence to suggest a deep-seeded connection between mental health and exposure and affiliation with the natural environment. Lastly, I present the complementary and cumulative mental and physical health co-benefits of accessing GI for physical activity.

In studies measuring the effect of exposure to nature and a variety of mental health outcomes—but not always explicitly attributing these outcomes to stress, affect, or cognition—it has been found that exposure to nature brings with it a variety of improvements to mental health. Many studies set the natural environment against the urban built environment to test these relationships. It is hypothesized that modern tendencies towards the habitation of urban environments where green (and blue) infrastructure has been erased or diminished is at odds with our innate positive predisposition to the natural environment. With the first cities sprouting up in Mesopotamia a mere (evolutionary speaking) 6,000 years ago or so, humans are thought to have not inhabited artifact-dominated environments long enough to have developed any positive predisposition to them or the ability to fully navigate the "new" challenges these environments pose to our mental health. This is not to say that urban environments need to be taxing to mental health. Humans and human evolution are dynamic, and we have certainly adapted quite rapidly to our own urban creations, but we have not outgrown our need for nature. What this is to say is that the "discord" or mismatch between what the body and mind need and what urban built environments often fail to provide, exposure to nature, has proven detrimental to our mental health.

Measuring general psychiatric morbidity and controlling for a number of confounding factors, those living in rural areas with greater exposure to nature have been found to have a much lower prevalence of psychiatric morbidity as compared to persons living in cities (Lewis & Booth, 1994). This should not lead us to conclude that persons living in cities should move to rural environments to reduce psychiatric morbidity. Increasing exposure to nature by including GI in the city can also bring about these benefits. Lewis and Booth (1994) also found that people living in built-up areas with access to gardens and other greenspaces had a lower prevalence of psychiatric morbidity as compared to people in built-up areas with no such access. Studies such as this reveal general associations between greater exposure to the natural environment and improved mental health, and this has created a much needed basis on which to build and address some obvious limitations to concluding clear cause and effect.

Some of the latest findings employ a study design that accounts for individual characteristics, such as personality, which might lead one to self-select to be nearer to greenspace (White, Alcock, Wheeler, & Depledge, 2013). Self-selection, in this case, would occur if individuals that chose to live near greenspace had personality traits associated with positive mental health. This would mask any benefit of exposure to nature as a treatment or greenspace causing improved mental health. When controlling for self-selection, there were small effects of more greenspace in urban areas resulting in lower mental distress. In addition to the small effects in mental distress or poor mental health, this study also claims to be the first to demonstrate "a significant association between greenspace and psychological health using a positive, evaluative index of well-being" (p. 7). Again, ruling out self-selection, greater exposure brought about the same small but cumulatively significant effect on overall well-being.

Prospective studies also get us closer to cause and effect. Recent longitudinal evidence has revealed that GI in one's environment is indeed important in delivering sustained gains in general mental health (Alcock & White, 2014). People who moved to environments with more GI had immediate improvements in mental health, and this positive change was sustained over many years. The prospective nature of this study suggests that greenspace does indeed have a sustained effect on baseline mental health, supporting a "shifting baseline" hypothesis. The baseline hypothesis was supported to a greater degree than an alternative "adaptation" hypothesis (initial benefits that dissipate over time and end in a return to a previous state) or a "sensitization" hypothesis (little immediate benefits but longer term improvements to mental health). Again, moving to greener countryside environments is not necessarily the answer. Those moving to greener urban environments also showed improved mental health. Strangely enough, the converse was not necessarily true. Those who moved to less green areas experienced decreased mental health before the move, but mental health returned to a normal state after the move in somewhat of an adaptive fashion. It could be that the anticipation of the move is what reduced mental health or that diminished mental health state is what precipitated the move (although the authors discount this second possibility as unlikely considering move motivations were similar across the sample). This is yet unexplained, but it is possible that the benefits that have accrued in the greener environment before the move dip in anticipation of losing them but are then restored, in a protective fashion, for a period of time even after moving to a less green environment.

In another rigorous study that examined the relationship between neighborhood greenness and mental health, it was found that the greenness of one's environment was a significant independent predictor of improved mental health even when accounting for physical activity (recreational walking) and social cohesion (discussed in a subsequent section) (Sugiyama, Leslie, Giles-Corti, & Owen, 2008). As stated in Chapter 6 on physical activity, mental health, physical activity, and social capital are the "big three" that have garnered the majority of the attention in research examining the relationship between health and the built and natural environments. What this study allows us to conclude through its examination of all three domains is that perceived improvements in general mental health may be attributed to the restoration gained by passive exposure to greenness, independent of exposure gained by actively accessing it for physical activity and also independent of the mental health benefits of the social interaction that can take place in it. The authors attribute ART as one explanation of this relationship independent of physical activity and social capital, but, as is common, there was no clarification as to how attention restoration brings about these gains in general mental health. Although stress recovery is also cited without any accompanying explanation as to how it influences mental health, unlike ART there is a larger body of extant literature supporting the connection between exposure to the natural environment, stress, and a variety of health outcomes.

Stress

Stress has become a part of the everyday lexicon and being "stressed out" a part of many people's everyday life. The health effects of the body's physiological response to chronic stress include an increased risk of anxiety, depression, digestive problems, heart disease, sleep problems, weight gain, memory and concentration impairment, and compromised immune system functioning (Mayo Clinic Staff, 2013). As we can see from the litany of the effects of stress, the body's physiological response to stress can lead to mental (e.g. depression) and physical (e.g. heart disease) ills, and it can also interrupt other activities needed to maintain health such as adequate sleep and cognitive functioning. Everyday exposure to GI is more likely to have the greatest benefit to reducing the maladies caused by chronic stress, but it may also be beneficial to the recovery from acutely stressful episodes. It is possible that the overall reduction in stress brought about by exposure to nature could make acutely stressful episodes less severe. Reducing the severity of acutely stressful episodes has implications for health. For example, among those already suffering from heart disease and weakened cardiac functioning, some *acutely* psychophysiological (emotional) stressful circumstances can lead to heart attacks, arrthymias, and even death (Vlastelica, 2008; Zipes & Wellens, 1998). In addition to recovery from chronic or acutely stressful experiences, GI may also mitigate stressful noise and visual cues (e.g. blight, traffic noise) (Hartig, Mitchell, de Vries, & Frumkin, 2014). As we will see, exposure to the natural environment can take many forms that range from simply viewing images of nature to actively accessing GI.

The psychophysiological or psychoevolutionary stress recovery theory (Ulrich, 1983) posits that our innate connection with the natural environment results in a fairly rapid reduction in stress when viewing natural elements or pleasing natural landscapes. With this theory as a basis, but not measuring stress directly, Ulrich's (1984) study of hospital patients' window views of nature has been continuously cited as evidence of the connection between exposure to nature and improved health. This study into the therapeutic properties of nature views revealed that hospital patients recovering from surgery had shorter hospital stays, lower intake of potent narcotic pain drugs, and more favorable evaluations by nurses if their hospital room windows allowed views of trees rather than views of a brick wall (Figures 7.2a–b). Even though this study has been criticized as having serious limitations in its sample size, data collection period, and outcome measures (see Health Council of the Netherlands and Dutch Advisory Council for Research on Spatial Planning, 2004, p. 41. Note this report incorrectly cites the date of Ulrich's 1984 article as 1983), it nonetheless spurred a considerable amount of research that seems to confirm the broader findings that exposure to nature has health benefits. Ulrich (1984) also cites the ability of nature to "sustain interest and attention" (i.e. ART) as a theoretical basis, although he appears to abandon this theory in his later work that focused solely on the stress mechanism. Nonetheless, in this study ART is a viable alternative or complementary explanation. In line with ART, viewing nature through the window would allow recovery from attention fatigue. Thus, the ability to concentrate would be improved. It could be this improvement to cognition that allowed patients to process information more accurately leading to shorter recovery durations and more favorable evaluations. In health care settings in particular, a patient's ability to process information and instruction given by health care providers is critical to enhancing care and health outcomes.

Years later, Ulrich and colleagues targeted stress recovery more directly and focused on passive exposure to nature not through a window but by comparing the effect of videotaped nature views to videotaped urban views largely devoid of natural elements (Ulrich et al., 1991). It was the first such study to test the psychophysiological stress recovery theory using a number of objective physiological indicators. Using muscle tension, skin conductance, and pulse transit time which correlates with systolic blood pressure, it was found that "recovery from stress was much faster and more complete when subjects were exposed to the natural settings" as opposed to the urban settings (Ulrich et al., 1991, p. 218). It was not necessary to view nature first-hand; images of nature alone induced stress recovery. Also using a number of physiological measures of stress (electroencephalography (EEG), electromyography (EMG), blood pressure volume, and a validated state-anxiety survey), another study returned to the window view again in the potentially stress-inducing environment of the workplace, and it also tested whether bringing nature indoors has any added benefit. It was found that viewing nature through a window was the most effective way to reduce stress and anxiety, more so than having a view of the city or solely having more immediate contact with nature in the form of indoor plants (Chang & Chen, 2005).[2] There was an

Figure 7.2a–b: Opposing hospital patient vistas

Source: a) Grid Engine, b) Todd Petit.

added benefit of indoor plants, with the greatest benefits in reducing state-anxiety experienced by employees who had both window views of nature *and* indoor plants. Greater exposure to nature = more complete stress recovery.

Viewing nature not just through a window of a building but also through the windscreen of an automobile also has mental health benefits. Research demonstrating greater stress recovery in natural versus largely artifact-filled environments has also been found to occur while driving (Parsons, Tassinary, Ulrich, Hebl, & Grossman-Alexander, 1998), an activity that can be highly stressful and one that a greater proportion of the world's population is performing every year. In the last decade there has been an encouraging decrease in passenger vehicle travel volumes in high-income economies (Van-Dender & Clever, 2013), but the total number of cars and trucks registered worldwide has continued to climb. There are now over one billion cars and trucks registered worldwide with some of the steepest climbs occurring in China and India, countries with enormous populations and the greatest economic potential to maintain these trends (Sousanis, 2011). Incorporating GI into roadway design can reduce the mental health consequences of this increasingly common activity. Not only do artifact-dominated drives elevate blood pressure and skin conductance—relative to nature-dominated drives—but, once one is stressed, a return to baseline levels of heart rate and skin conductance are more likely among those driving on roadways where nature was visible. While driving, having views of nature resulted in comparatively lower levels of stress—perhaps suggesting a coping mechanism—but viewing nature also aided the ability to recover from stress. Even though the population in the windscreen study was unique (college students) and the results have not been replicated, the experimental design of this research and the global trends in auto use should lead to a serious consideration of incorporating nature in roadway design to reduce the stressfulness of driving and increase the ability to recover from this increasingly common activity.

A number of cross-sectional studies have shown that when exposure to nature is achieved through accessing GI—not just viewing images or having a window view—increased access is accompanied by reduced levels of stress. Those who report visiting greenspace more frequently and spending more time in greenspace reported fewer stress-related illnesses (Grahn & Stigsdotter, 2003), and those who do not report being stressed are much more likely to visit greenspaces at least a few days a week (Stigsdotter et al., 2010). Although in no way causal, this does coincide with the stated reasons greenspace users provide as benefits of visiting greenspace.

Stressed individuals cite "reducing stress and relaxing" and "obtaining peace and quiet without noise" more frequently than non-stressed individuals. Noting that the closer one lives to greenspace the less often they suffer from stress (Grahn & Stigsdotter, 2003), the incorporation of GI systems that increase access would provide this self-identified resource for reducing stress. Among older individuals, the belief in the health benefits of parks corresponds to their stress-reducing power. Older residents in high versus low stressed groups stay in parks longer and had lower blood pressure if they believed that there were health benefits to their park visit (Orsega-Smith & Mowen, 2004). Also, for residents in public housing with nearby GI, this amenity allowed residents to better cope with the management, duration, and severity of major issues that have the potential to be stressful (Kuo, 2001).

This evidence is complemented by an examination of the moderating effect of the amount of greenspace on the relationship between stress and overall health. The relationships between stressful life events and the number of health complaints and perceived general health are significantly moderated by amount of greenspace within a 3km radius of home; those with a greater amount of greenspace are less affected by a stressful life event than those with less greenspace (van den Berg, Maas, Verheij, & Groenewegen, 2010). Having a greater amount of greenspace close to home increases the potential for exposure gained through access, but it does not guarantee it, but, even if direct contact through access cannot be confirmed, there is certainly a higher likelihood of increased involuntary exposure to nature in the process of performing everyday tasks.

More greenspace appears to be beneficial to reducing stress, but there also appears to be such a thing as too much "greenness." An experimental study of the effect of street tree density (through video exposure) on stress recovery revealed that there are diminishing returns in stress recovery with high levels of tree density (Jiang, Chang, & Sullivan, 2014). Varying the level of street tree density had no significant effect on stress recovery in women. Men did experience stress recovery benefits with exposure to moderate levels of street tree density, but these benefits diminished as street tree density increased. Results such as this speak to the importance of GI design and should be encouraging to those trying to strike the right balance between the built and natural environments. Unadulterated views of nature are not necessary to deliver stress recovery benefits. A balance between the built and natural environment may actually be optimal for the delivery of some mental health benefits.

A number of studies in Japan of the effects of Shinrin-yoku, or actively accessing greenspaces for "forest bathing," found that blood pressure, heart rate variability, and salivary cortisol concentration were reduced after forest bathing with additional improvements in subjective feelings of being "calm," "comfortable," and "refreshed" (Tsunetsugu et al., 2007). Engaging the natural environment has positive effects on both objective and subjective measures of stress. Replicated years later with a much larger sample that compared the effect of the forest versus urban environment on physiological measures of stress (Park, Tsunetsugu, & Kasetani, 2010), forest bathing was found to reduce stress as measured by lower cortisol levels, lower pulse rate, lower blood pressure, greater parasympathetic nerve activity (indicative of rest), and lower sympathetic nerve activity (indicative of stress) as compared to urban environments.

Accepting that individual constraints (e.g. perceived safety issues, reduced mobility) and physical environment constraints (e.g. lack of nearby GI) will not make accessing GI possible for everyone, it should be encouraging that it is not necessary to access GI to receive stress recovery benefits. As stated before, stress-reducing benefits occur from exposure gained through access but also in a room with a view of nature (Hartig et al., 2003). In an increasingly urban world, this has serious implications for the mental health of a growing majority of the world's population with diminished opportunities for direct access to nature either because more time is spent indoors or because there is a lack of greenspace in one's immediate environment. Until GI pervades the urban environment, city dwellers will have a diminished capacity to affiliate with nature and benefit from its stress recovery properties, properties that not only reduce stress but allow one an increased capacity to better cope with the battery of potential stressors posed by the demands of the urban environment. Conserving nature in one's own backyard or personal space (Nielsen & Hansen, 2007) or viewing images of nature could be considered as a compromise solution, but this compromise to creating and conserving interconnected systems of GI would fail to deliver the myriad other health benefits of GI.

This section has revealed that exposure to the natural environment—exposure gained either by viewing or accessing it—can aid in stress recovery. The next section reveals the ability of the GI to ameliorate negative affect, but one must keep in mind that stress and affective state are intertwined. There is a large volume of evidence linking prolonged stress and stressful life events to negative emotions such as anxiety, hostility, and depression (e.g. Bolger, DeLongis, Kessler,

& Schilling, 1989). Emerging evidence suggests that there is a physiological explanation for stress influencing emotions. Stress-induced anxiety has been shown in mice models to be caused by the migration of immune system cells to the brain (Wohleb, Powell, Godbout, & Sheridan, 2013). Stress can influence affect, and new theories such as this help us understand how this occurs, but stress is not the only explanation for negative affect. We all experience stress, but not everyone experiences mood disorders such as chronic anxiety and depression. Some people are more susceptible, either biologically or through socially learned coping, to these disorders. Therefore, it is appropriate to examine how GI can promote positive affect and alleviate affective disorders separate from stress.

Affect

Affect is "any state that represents how an object or situation impacts a person" (Duncan & Barrett, 2007, p. 2). Affective state is represented in innate human feelings such as fear, anger, and joy which are then filtered through our learned cultural norms and expressed in emotions (Kelly, 2012) such as anxiety, depression, aggression, and happiness. Essentially, to be human is to have feelings, which are a product of our biologically determined affect, but how these feelings are expressed in emotions is determined by both biological factors and by social learning. The focus here is not on how biological and social factors influence the expression of emotions but rather on how GI can influence emotional states that are associated with morbidity (e.g. depression). This gets a bit tricky in that the terms affect, emotion, and mood are not used consistently and are, at times, used interchangeably. What is generally agreed upon is that mood and emotion represent different durations of feelings.[3] While this is important to note, the varied use of these terms is not detrimental to our understanding of the influence of GI on affect and our ability to conclude that exposure to nature has a positive influence on emotions and mood.

What is also important to note before proceeding further is that affect has been isolated here because there is a sufficient volume of literature connecting GI to affect and affect to health, but isolating it is not to imply that it operates independently from stress and attention restoration. Affect is isolated here because one can experience positive affect from exposure to GI without being stressed or attentionally fatigued and therefore in need of restoration, but when restoration does occur, it appears that positive affect joins the behavioral, cognitive, physiological, and social benefits that ensue from restoration.

The overwhelming majority of the research examining the role of affect on health has focused on how negative affective states increase physical and psychological morbidity. What is largely underdeveloped is our understanding of how positive affect can improve health (Cohen & Pressman, 2006; Pressman & Cohen, 2005) with positive affect reflecting a level of pleasurable engagement with the environment (Clark, Watson, & Leeka, 1989). Increasing the pleasurable engagement with the environment can cultivate the positive emotions that can counteract health issues rooted in negative emotions such as anxiety, depression, aggression, suppressed immune functioning, suicide, and even some cancers (Fredrickson, 2000). Positive affect has also been examined for its effect on one of the leading causes of mortality in higher income countries: coronary heart disease. Over the course of a decade, it was found that for every one-point increase on a five-point scale of positive affect (happiness, joy, contentment, enthusiasm), the rate of heart disease dropped by 22 percent (Davidson, Mostofsky, & Whang, 2010).

Before Ulrich's focus on stress, and to the best of my knowledge one of the first studies into the psychological benefits of exposure to nature, it was found that viewing nature scenes brought about an increase in positive affect as measured by affection, friendliness, playfulness, and elation (Ulrich, 1979). This study also revealed that viewing nature can reduce fear, which shows up again five years later in Ulrich's often-cited work on window nature views and patient recovery (see Stress section above). Reducing the patient emotion of fear is a complementary explanation to stress reduction and attention recovery that was shown to be associated with improved patient outcomes.

Similar to stress research, and at times folded into the same studies, the effect of greenspace on emotions and mood has been measured by comparing exposure to natural environments versus more urban and artifact-dominated environments with little or no greenspace. Although now firmly rooted in stress recovery, Ulrich et al. (1991) took into account affective states in addition to stress to measure restoration. After viewing distressing images, it was found that subjects that viewed natural scenes reported less fear and anger/aggression and much higher positive affect. This "recovery associated with the natural exposures was so pronounced in terms of the Fear, and especially the Anger/Agression [*sic*] and Positive affects factors, that post-recovery affective states were somewhat more positively-toned than those reported during the base-line period" (Ulrich et al., 1991, p. 220).

Startling. Not only did viewing images of nature produce affective recovery after being distressed, but subjects were better off in terms of affect than before they were distressed (i.e. better off than before they participated in the study).

Controlled simulations that compare exposure to natural environments versus urban environments have consistently revealed more positive emotional self-reports by those viewing natural environment scenes including higher levels of overall happiness and reduced anger and aggression (Hartig, Book, Garvill, Olsson, & Garling, 1996). Simulations are wonderful in that they allow for the control of potential confounding variables, but they cannot account for the myriad other senses, in addition to the visual, that may have an influence on one's interpretation and response to the environment. Complementing controlled simulations are field studies that engage all the senses in testing the nature/urban dichotomy. In one such field experiment, persons first completed tasks that demanded focused attention and then took a walk in either natural or urban environments (Hartig et al., 2003). While no improvement was found in overall happiness when walking in the natural environment, there was an increase in overall positive affect, and anger/aggression decreased relative to the pretest. The opposite pattern occurred in the urban environment. The same result of higher positive affect and happiness and lower anger and aggression was evident in an earlier field study that compared the effects of walking in natural versus urban environments (Hartig, Mang, & Evans, 1991). It is true that there are mental health benefits of physical activity (discussed shortly in the Mental Health Co-benefits section) and that these mental health benefits could be produced simply by walking itself, but this study revealed that these benefits are determined by the type of environment one takes a walk in. Only GI produced positive affect.

The most recent work measuring differential affective responses to natural versus urban environments has employed electroencephalogram (EEG) and EEG-based emotional recognition software to measure changes in emotions while navigating urban environments on a walk. It was found that pedestrians that were walking on a busy urban shopping street and then entered greenspace experienced reductions in arousal (long-term excitement) and frustration (Aspinall, Mavros, Coyne, & Roe, 2013). This methodological advancement allows for a confirmation of previous self-reports of positive emotional state and calmness associated with exposure to nature. The study design also accounted for inter-group variation which could be attributed to the characteristics of subjects in urban *or* natural environment groups. Subjects were not exposed to only one or the other type of environment but rather the same individuals transitioned between the two different environments.

People appear to be aware of the ability of the natural environment to restore affect. This has been confirmed when examining the mediating role of affective restoration in environmental preference (van den Berg, Koole, & van der Wulp, 2003). Not only was improvement in mood greater when exposed to videotaped simulations (with sound) of a walk in a natural versus an urban environment, but persons in need of restoration stated a significant preference for natural environments. This preference is much more than simply an idyllic and naive view of a peaceful and quaint rural life surrounded by nature. The preference for affiliating with nature is linked to people seeking it out for its restorative properties (van den Berg, Hartig, & Staats, 2007). Our preference for natural environments may stem from an innate evolutionary heritage as proposed in the biophilia hypothesis and others (e.g. Orians & Heerwagen, 1992) or may be socially learned, but, either way, and taken together with other nature versus urban findings, the human preference for nature and affective restoration (and other forms of restoration) appear to coincide.

One example of how GI might address a public health concern tied to the emotion of aggression is in children's bullying behavior. A lack of exposure to GI in children's lives could be a contributing factor to bullying behavior and its enduring physical and mental effects. Remedying this by providing the green outdoor environments, that have been shown to promote reduced BMI in children, may also provide the co-benefit of reduced aggression. It has been found that the least amount of bullying occurred in children's playspaces with a highly interactive and an engaging natural environment where exploration was encouraged (Malone & Tranter, 2003). This does not allow us to conclude that the engagement with the natural environment is what caused less bullying, but, similar to other emerging research topics, it provides an important basis to continue to test the importance of natural playspaces and engagement with the natural environment in encouraging more positive emotions in children.

Greening might also be an important environmental treatment to reduce one of the most aggressive of behaviors, gun violence. In Philadelphia, the greening of vacant lots was associated with a reduction in gun assaults over a 10-year period (Branas et al., 2011). This highly aggressive and deadly behavior appears to be reduced by improving the quality of GI in neighborhoods, but it could be that improvements to GI were simply an indicator that someone cared and investments in neighborhoods were being made. It could very well be that putting in sidewalks could have the same effect, but operating under the theory supported in other studies that there truly is something unique about GI leading to more positive emotions, the connection of GI to gun violence does seem plausible. Of course, there are currently people killing one another in parts of the world with abundant GI, so there is obviously some threshold of other needs that need to be met in order for GI to have an effect on creating a more peaceful coexistence.

Continuing along the lifecourse continuum, exposure to the natural environment appears to also benefit affective states of older persons in institutional settings with degenerative mental disorders. Among those with late stage dementia in nursing homes, natural sounds and pictures have been found to reduce agitation but not necessarily aggression (Whall, Black, & Groh, 1997). Furthermore, the risk of developing dementia has been found to be dramatically reduced with exposure to and affiliation with nature achieved with regular gardening (Fabrigoule et al., 1995). While aggression was not reduced in dementia patients when representations of nature were brought into nursing homes, the more complete opportunity for immersion in nature brought about by the installation of gardens was found to reduce incidents of aggressive behavior in Alzheimer's patients when compared to facilities without this resource (Mooney & Nicell, 1992).

Nature-based walks have been found to be beneficial to those diagnosed with mental disorders. In a group of individuals with major depressive disorder, positive affect improved to a greater extent in nature-dominated walks as compared to urban walks (Berman et al., 2012). The authors noted that the effect sizes were nearly five times larger in this group than that had been previously found among healthy individuals. While the act of walking in both urban and natural surroundings reduced negative and improved positive affect, only the nature walk benefited positive affect significantly when compared to the urban walk. This study also tested the interactive effects of memory and mood and concluded that these were distinct processes in this population. While improvements in affect do not appear to be accompanied by improvements in the cognitive process of memory in this population, affect and cognition are inseparable in order for humans to function and navigate environments.

Kelly (2012, p. 164) provides an example of how they are inseparable in stating:

the purely 'rational' understanding or knowledge of something by cognitive processes is not motivating without affect. Likewise, the purely 'emotional' originating from within the affect system operates in the dark without cognition. From the point of view of survival of the species, a real separation in the functioning of these two systems would be a disaster. For instance, the fear triggered … when a car is spotted heading right at you would not be acted on properly if you did not have knowledge about the possible consequences of not moving out of the way. You might simply stand there frozen in terror. By the same token, the knowledge that the car is coming at you is useless until fear motivates you to move. Otherwise, you might stand there wondering about the paint job or the design of the front bumper. The bottom line is that the cognitive system and affect system have an open channel to one another. They always operate together even though they have different functions.

Likewise, there is no neurobiological basis for a distinction between affect and cognition supporting the well-formulated argument for how the distinction is phenomenological and not ontological (Duncan & Barrett, 2007). Taking a step back, the point here is not to resolve any debate on the relationship between affect and cognition. The takeaway message is that they are inseparable for human functioning. Therefore, it is critical to also understand cognition in addition to affect (and stress). This is done with relative ease because cognition (i.e. attention) has often been measured separately from stress and affect in the literature exploring the mental health benefits of exposure to the natural environment.

Cognition and attention

An enduring definition of cognition is all processes by which "sensory input is transformed, reduced, elaborated, stored, recovered, and used" (Neisser, 1967, p. 4). Essentially, cognition is what most would consider thinking or the process of receiving information, processing it, and then applying it to make decisions. Tasks that require sustained and voluntary directed attention can be mentally draining, but so too can the actions associated with navigating the everyday environment. Everyone is prone to the mental fatigue that results from meeting everyday demands. Recovery from this fatigue can take many forms. Active meditation or a simple "time out" are techniques that relieve us from the effort of cognitive engagement, but what has been found is that we do not necessarily need to separate ourselves from environmental stimuli to recover from attention fatigue.

We may only need passive exposure to the natural environment to achieve this form of cognitive restoration (Berman, Jonides, & Kaplan, 2008). We revisit ART here as the theoretical justification for how exposure and access to GI can restore cognitive functioning and processes, such as inhibition, that are associated with health outcomes.

Attention restoration theory has great potential to help us connect the restorative properties of nature to psychological health, but the many studies that cite ART as the theoretical basis for the mental health benefits of exposure to the natural environment have not made it clear how recovery from attention fatigue brings about improved mental health. This is remedied somewhat in the psychological health domain in a conceptual framework outlining the comprehensive array of relationships between GI and health (Tzoulas et al., 2007), as shown in Figure 7.3. Here are *stress* and *emotion* (affect), discussed in the previous two sections, but also displayed is the relationship between *attention* and *cognition*. The slight exception I take with this representation is that, unlike stress and emotion, attention likely does not stand alone as separate from stress and cognition. In order to have a measurable effect on health, depleted attention would need to cause stress or hinder cognition, both of which could also affect emotion. Since attention can be compromised independent of stress, in these cases the only clear pathway by which we can link exposure to GI to improved attention and then to mental health benefits is through cognition, recognizing that cognition has an influence

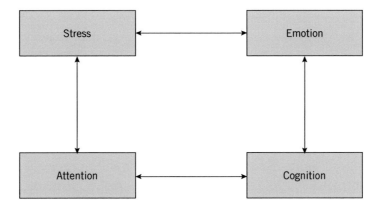

Figure 7.3: Relationship between domains of mental health
Source: Adapted from Tzoulas et al. (2007).

on health behaviors and outcomes. This is why attention and cognition are taken together here.

Some types of environments demand more of the attention that can be fatigued over time but that can then be restored by exposure to the natural environment. The previously cited study by Aspinall et al. (2013) that employed EEG to measure affect also measured directed attention. While the restorative move from urban to green environment was found to bring about "a greater range and subtlety of emotional response," the opposite movement from green to the more attentionally demanding urban environment with heavy traffic and more people brought about a clear effect in engagement and alertness and directed attention. The urban environment requires greater directed attention that can cause attention fatigue, and the natural environment provides the opportunity to recover from this fatigue with resultant cognitive improvements. After completing a mentally taxing task, video exposure to the natural environment reduced heart rate, and subjects performed better on tasks that demanded directed attention as compared to those who viewed urban settings largely devoid of nature (Laumann, Gärling, & Stormark, 2003). This heightened cognition brought about by the restorative effects of nature was also experienced in first-hand engagement with the natural environment. Subjects performed better on proofreading tasks after walking in nature as compared to walking in urban environments (Hartig et al., 1991). Whether viewing images of nature or experiencing it first-hand, exposure to the natural environment was accompanied by restored attention and subsequent improvements in cognition.

We again look to children to understand how the ability of nature to improve attention and cognition might be applied to health. Since exposure to GI can restore attention, it may be able to ameliorate behavioral disorders exacerbated by mental fatigue. One such disorder is Attention Deficit Hyperactivity Disorder (ADHD), the most common behavioral disorder among children in the USA. The symptoms of ADHD overlap with the symptoms of mental fatigue (e.g. distracta-bility, irritability) caused by sustained and depleted directed attention. While there may be some merit in the relationship between a decrease in children's exposure to nature and an increase in disorders such as ADHD, what is clear is that symptoms of mental fatigue and ADHD overlap and the restorative properties of exposure to GI can alleviate these symptoms. Indeed, when parents were asked about the "aftereffects" of green outdoor activity on their children's ADHD symptoms, it was found that green outdoor activity had a significantly greater effect than urban outdoor and indoor activities in reducing ADHD symptoms (Kuo & Faber Taylor, 2004). It was not activity alone, but activity in GI that brought about a reduction in ADHD symptoms. In fact, when children were in groups, indoor activity actually increased ADHD symptoms while outdoor green activity was the only setting that significantly reduced symptoms. On a personal note, I have grappled with ADHD my entire life, and now I may have an explanation for the feelings of contentment gained when I walk in a park or hike in the wilderness. Being somewhere between childhood and retirement age now, it is encouraging that these benefits appear to continue beyond childhood. The ability of nature to improve concentration has also been found among the elderly in geriatric care facilities (Ottosson & Grahn, 2005). I have a few years before my sunset in a geriatric care facility, but it is comforting to know that when I get there I will be able to concentrate on my Bingo after a walk in the park.

The ability of nature to restore attention has also been explored for its role in self-discipline that plays an essential part in the performance of many health behaviors. This has been tested among girls living in inner-city dwellings. The underlying hypothesis is that if nature can restore attention then it could improve concentration, reduce impulsive behavior, and support the delay of gratification associated with outcomes such as academic achievement, vandalism and violence, and possibly teenage pregnancies (Taylor, Kuo, & Sullivan, 2002). Measuring the amount of nature that could be viewed from windows of inner-city dwellings, it was found that all three aspects of girls' self-discipline (concentration, impulsive behavior, delay of gratification) showed a positive and significant relationship with the greenness of the immediate vicinity that could be viewed from the dwelling. There may also be benefits to the restorative properties of window views of nature among mothers in these same type of dwellings. Views of nature were found to reduce the psychological and physical violence aimed at the partners of mothers in inner-city dwellings (Kuo & Sullivan, 2001). Aggression was folded into the previous section covering affective states, but aggression has also been explored as an outcome of mental fatigue coinciding with the proposition that depleted directed attention can lead to irritability (Kaplan, 1995, p. 172). In Figure 7.3, attention and emotion are not connected, but maybe they should be.

Again, there is certainly overlap and interactions among stress, affect, and cognition, and it is the natural environment that can bring about improvements in all three aspects of mental health. While parsing out these interactions is an important academic exercise, this should not deter the practical application of increasing exposure to the natural environment in order to achieve the mental health benefits associated with reducing stress, increasing positive affect, and improving cognition. Quite frankly, a complete understanding of the psychological and physiological mechanisms elucidating how people achieve these benefits is not essential to improving people's lives. Rather, it is only essential that people experience these benefits, and it is quite clear that we know how to bring about these mental health benefits: access and exposure to GI.

While GI is supporting mental health, it is simultaneously providing the array of other health benefits discussed up to this point, some of which are essential to sustaining life, others to enhancing it, but it is not just the volume of co-benefits of GI that permeate life and support health that should be respected but also the interconnections between these co-benefits. For example, struggling to secure clean water can be stressful. Protecting GI could aid in the provision of clean water in some circumstances, and this could then reduce stress, but what is more interesting and potentially powerful is the *cumulative* health effects of GI on the co-benefits of providing water and reducing stress. One area where these cumulative effects have been studied is the mental health benefits of performing physical activity in GI.

Mental health co-benefits of green exercise

There are many simultaneous public health co-benefits occurring from the presence, access, and exposure to GI, but seldom is the potential for cumulative health benefits considered in research examining the nature and health relationship. The potential cumulative mental health benefits of performing physical activity in greenspace is one area where this shortcoming is being addressed. When a person accesses GI for physical activity, by default he/she is subject to the mental health benefits of exposure to nature. There are a number of mental health benefits of exposure to nature, and there are numerous mental health benefits of physical activity alone, but are there greater mental health benefits brought about by physical activity that includes exposure to nature?

Although it is not entirely clear how the physiological and psychological aspects of physical activity work together to produce mental health benefits, physical activity has been shown to have a positive effect on stress responsivity, mood state, self-esteem, depression, anxiety, premenstrual syndrome, and body image (Scully, Kremer, Meade, Graham, & Dudgeon, 1998). Physical activity performed indoors in gyms and on treadmills can produce these mental health benefits, but, as we have seen, a number of these benefits (e.g. reduced stress, improved mood) also occur simply from exposure to GI. In a meta-analysis of physical activity that includes exposure to the natural environment, it was found that "acute short-term exposures to facilitated green exercise improves both self-esteem and mood irrespective of duration, intensity, location, gender, age, and health status" (Barton & Pretty, 2010, p. 3950). Green exercise, especially exposure to water features, is good for people with varied individual characteristics performing a variety of activities, but it is not clear how much of the positive contribution is from the physical activity or from exposure to the natural environment.

Taking the body of literature as a whole to parse this out, there has been found to be "promising effects on self-reported mental wellbeing immediately following exercise in nature which are not seen following the same exercise indoors" including "greater feelings of revitalization and positive engagement, decreases in tension, confusion, anger, and depression, and increased energy" (Thompson Coon et al., 2011, p. 1761). There appears to be an added mental health benefit to physical activity performed in GI, but this still leaves us with the challenge of getting people to maintain this behavior and make it a part of their daily lives.

There is some promise that exercise in the natural environment may not only lead to greater improvements to mental health than those achieved through activity alone, but exercise in this setting may also make it more likely that physical activity is sustained. People who perform activity in natural settings report "greater enjoyment and satisfaction" with the activity and "a greater intent to repeat the activity at a later date" (Thompson Coon et al., 2011, p. 1761). This intent has been confirmed with evidence showing that performing activity in more mentally restorative outdoor settings was more predictive of exercise frequency than activity performed in less restorative indoor settings (Hug, Hartig, Hansmann, Seeland, & Hornung, 2009). Of course, issues of safety and climate that influence one's decision to perform activity outdoors cannot be overlooked, but, if overcome, people will gain greater mental health benefits and gain them more often than if they were to perform the same activity indoors in settings devoid of nature.

Taking a closer look at some of the individual studies that led to Thompson Coon et al.'s (2011) conclusion drawn from the overall body of evidence, when measuring the mental health effects of 10 green exercise activities including walking, cycling, horse riding, fishing, canal boating, and conservation activities, it was found that participants had markedly improved self-esteem and were significantly less angry, depressed, confused, and tense after engaging in green exercise (Pretty et al., 2007). Some of these activities likely do not demand the levels of exertion necessary to meet the recommended levels necessary to achieve maximum physical health benefits, but their performance in a natural setting led to measurable mental health benefits (Figure 7.4). This is good news in that a variety of activities performed in GI can accommodate individual preferences.

Again, getting at whether the greenness of the activity can bring about greater benefits than activity alone, there are a number of studies that compare mental health outcomes of performing physical activity in predominately natural versus urban settings. Among a small sample of runners, it was found that running in a park had a measured effect on reduced anxiety and depression when compared to running in an urban setting, even an urban setting that was relatively "benign" (it had some greenery and was of relative low density) (Bodin & Hartig, 2003).

Figure 7.4: Pine forest restoration

Source: US Forest Service.

Note: Conservation activities can be both physically demanding and mentally restorative. They also have the added benefit of protecting the GI that supports health in these and many other ways.

Even images of nature have been shown to have positive effects with marked improvements to self-esteem when physical activity was performed viewing both rural and urban "pleasant" settings as compared to exercise alone (Pretty, Peacock, Sellens, & Griffin, 2005). Urban pleasant settings were those with natural elements and rural pleasant settings contained pastoral elements without evidence of human degradation. Both urban unpleasant scenes devoid of nature and rural unpleasant scenes that included human degradation of the natural environment showed the opposite effect, a reduction in the positive effect exercise had on self-esteem. It is very interesting to note that "rural unpleasant scenes had the most dramatic effect, depressing the beneficial effects of exercise on three different measures of mood" (Pretty et al., 2005, p. 319). Threats to the countryside evidenced by human inter-ference and degradation had a greater negative effect on mood than unpleasant urban settings. This is potentially very telling in that it could reflect an assault on a biophilic tendency where people are saddened more by GI being degraded than by perceived degradation of the human-made environment where nature is presumed to be already lost.

The mental health benefits of physical activity in different types of environ-ments have also been examined using both negative (psychiatric morbidity) and positive (mental well-being) mental health outcomes. Physical activity performed in the natural environment was associated with a greater reduction in the risk of psychiatric morbidity than physical activity performed in other types of environ-ments (Mitchell, 2012). Woods and forests showed a particularly significant effect on reducing psychiatric morbidity as compared to other forms of green and non-green environments, and this was also the only form of GI that held a positive association with positive mental well-being, albeit only marginally. Other forms of settings for physical activity (e.g. gyms, local streets, sports fields) were associated with positive mental well-being but not a reduction in psychiatric morbidity. This confirms, at the population level, that physical activity in many types of environ-ments has positive mental health benefits and opens up the proposition that certain types of GI (e.g. woods and forests) may be better suited to reducing the risk of poor mental health.

The evidence strongly suggests that performing physical activity in natural settings improves mental health, with greater gains in mental health than could be achieved from either physical activity or exposure to nature alone. There is enough evidence demonstrating these cumulative benefits that health-promotion practice has implemented green activity to treat mental health disorders such as depression. *Ecotherapy* has been supported with many millions of dollars in the UK as a form of free and accessible treatment for those under mental distress (Hine, 2007). Ecotherapy activities include gardening, farming, art and craft, exercise, or environmental conservation work (e.g. Green Gyms, discussed in Chapter 6 on physical activity). In a survey of those participating in ecotherapy activities, an overwhelming majority reported positive effects in their physical and mental health. There will of course be situations where pharmaceutical treatment and its accompanying side effects are necessary to treat mental health conditions, but ecotherapy may be an effective complementary treatment for those with mild, moderate, and even serious mental disorders (recall the positive effect of nature walks among those with major depressive disorder (Berman et al., 2012)). The GI needed to perform ecotherapy also supports the many other health benefits outlined in this book. Achieving these benefits while performing physical environmental conservation work is of particular ecological importance as it is an act of ecological feedback where physical and mental health benefits are achieved while protecting the very GI necessary for our continued health and survival. The fact that green activities, such as outdoor conservation work, are often done in groups also has the added health benefit of social interaction. The promise of GI in promoting social interaction and social capital essential for health is discussed in the following chapter.

Summary

Ecological public health interventions that increase exposure and access to GI are necessary to unlock the restorative potential of the environment. Greenspaces are an "escape facility" (Chu, Thorne, & Guite, 2004) that facilitate mental restoration. This restoration is expressed in reduced stress, heightened positive affect, and improved cognitive functioning. Recovery from stress has been the focus of most of the research into the restorative properties of exposure to the natural environment with the consistent conclusion being that exposure to nature is a powerful stress-reducing treatment for a diverse cross-section of people. The focus on stress does not diminish the potential health benefits of heightened positive affect and improved cognitive functioning. These outcomes themselves are associated with positive mental health conditions and behaviors, and they interact with each other and with stress responses. The increased happiness, reduced anger, and improved cognition brought about by exposure to GI, combined with its stress-reducing properties, together reveal the power of nature to both treat and prevent many mental health conditions and their physical manifestations. Particularly in rapidly expanding and mentally taxing urban environments around the globe, urban nature is essential to fulfill a deep-seeded need for mental restoration, a need people fulfill by seeking out nature.

There are also cumulative mental health benefits when exposure is achieved by accessing GI for physical activity. There are well-documented mental health benefits to physical activity alone, but these benefits are more pronounced when activity is performed in GI. Further exploration of the many co-benefits of exposure and affiliation with the natural environment is exactly what is needed to paint a truer representation of nature's sweeping role in supporting health. It is quite possible that the GI that improves water and air quality, reduces the incidence of infectious disease, helps to mitigate climate change, promotes physical activity, and so on has a greater positive effect on physical, mental, social, and spiritual well-being when the cumulative effect of these ecosystem services are considered together.[4]

Notes

1. For a detailed, although a little dated, summary of the theoretical approaches taken in the study of the psychological benefits of nature as well as a literature review see Rohde and Kendle (1994).
2. There are a number of studies examining the benefits of bringing nature into indoor environments with no negative but mixed positive results (Bringslimark, Hartig, & Patil, 2009). Since the focus of this book is on GI, the role of indoor plants is not explored in depth here.
3. Mood is often used to refer to longer duration emotional state, and emotion is often used to refer to a short-term immediate affective response. When citing the evidence to come, we have stated the specific emotion (e.g. happiness, anger) when possible instead of simply noting positive or negative changes in mood.
4. Research into this connection between the natural environment and mental health has typically been conducted in regions of the world where the perils of the wilderness have been tamed and where the procurement of the ecosystem services that support life are no longer an immediate and constant preoccupation. In these regions where gray infrastructure and complex social and economic systems have been designed to deliver nature's bounty, it is assumed

that when we turn on the tap, the water will flow, and when we go to the grocer, food will be there. The limitation to concluding a universal psychological benefit to exposure to nature is a lack of evidence from regions of the world where mere survival in the natural environment is still a daily preoccupation and where varied cultural attitudes towards nature could translate into varying trained psychological responses.

References

Alcock, I., & White, M. (2014). Longitudinal effects on mental health of moving to greener and less green urban areas. *Environmental Science & Technology, 48*(2), 1247–55.

Appleton, J. (1975). *The experience of landscape*. London: Wiley.

Aspinall, P., Mavros, P., Coyne, R., & Roe, J. (2013). The urban brain: Analysing outdoor physical activity with mobile EEG. *British Journal of Sports Medicine*, online, 1–6.

Barton, J., & Pretty, J. (2010). What is the best dose of nature and green exercise for improving mental health? A multi-study analysis. *Environmental Science & Technology, 44*(10), 3947–55.

Berman, M. G., Jonides, J., & Kaplan, S. (2008). The cognitive benefits of interacting with nature. *Psychological Science, 19*(12), 1207–12.

Berman, M. G., Kross, E., Krpan, K. M., Askren, M. K., Burson, A., Deldin, P. J., et al. (2012). Interacting with nature improves cognition and affect for individuals with depression. *Journal of Affective Disorders, 140*(3), 300–5.

Bird, W. (2007). *Natural thinking: Investigating the links between the natural environment, biodiversity and mental health*. Sandy, UK: Royal Society for the Protection of Birds.

Bodin, M., & Hartig, T. (2003). Does the outdoor environment matter for psychological restoration gained through running? *Psychology of Sport & Exercise, 4*(2), 141–53.

Bolger, N., DeLongis, A., Kessler, R. C., & Schilling, E. A. (1989). Effects of daily stress on negative mood. *Journal of Personality and Social Psychology, 57*(5), 808–18.

Branas, C. C., Cheney, R. A., MacDonald, J. M., Tam, V. W., Jackson, T. D., & Ten Have, T. R. (2011). A difference-in-differences analysis of health, safety, and greening vacant urban space. *American Journal of Epidemiology, 174*(11), 1296–306.

Bringslimark, T., Hartig, T., & Patil, G. G. (2009). The psychological benefits of indoor plants: A critical review of the experimental literature. *Journal of Environmental Psychology, 29*(4), 422–33.

Chang, C., & Chen, P. (2005). Human response to window views and indoor plants in the workplace. *HortScience, 40*(5), 1354–9.

Chu, A., Thorne, A., & Guite, H. (2004). The impact on mental well-being of the urban and physical environment: An assessment of the evidence. *Journal of Public Mental Health, 3*(2), 17–32.

Clark, L., Watson, D., & Leeka, J. (1989). Diurnal variation in the positive affects. *Motivation and Emotion, 13*(3), 205–34.

Cohen, S., & Pressman, S. D. (2006). Positive affect and health. *Current Directions in Psychological Science, 15*(3), 122–5.

Cohen, S., & Spacapan, S. (1978). The aftereffects of stress: An attentional interpretation. *Environmental Psychology and Nonverbal Behavior, 3*(1), 43–57.

Davidson, K. W., Mostofsky, E., & Whang, W. (2010). Don't worry, be happy: Positive affect and reduced 10-year incident coronary heart disease: The Canadian Nova Scotia Health Survey. *European Heart Journal, 31*(9), 1065–70.

de Vries, S., van Dillen, S. M. E., Groenewegen, P. P., & Spreeuwenberg, P. (2013). Streetscape greenery and health: Stress, social cohesion and physical activity as mediators. *Social Science & Medicine, 94*, 26–33.

Duncan, S., & Barrett, L. F. (2007). Affect is a form of cognition: A neurobiological analysis. *Cognition & Emotion, 21*(6), 1184–211.

Fabrigoule, C., Letenneur, L., Dartigues, J. F., Zarrouk, M., Commenges, D., & Barberger-Gateau, P. (1995). Social and leisure activities and risk of dementia: A prospective longitudinal study. *Journal of the American Geriatrics Society, 43*(5), 485–90.

Fredrickson, B. L. (2000). Cultivating positive emotions to optimize health and well-being. *Prevention & Treatment, 3*(1), online.

Grahn, P., & Stigsdotter, U. A. (2003). Landscape planning and stress. *Urban Forestry & Urban Greening, 2*(1), 1–18.

Grinde, B., & Patil, G. G. (2009). Biophila: Does visual contact with nature impact on health and well-being? *International Journal of Environmental Research and Public Health, 6*(9), 2332–43.

Hartig, T. (1993). Nature experience in transactional perspective. *Landscape and Urban Planning, 25*(1–2), 17–36.

Hartig, T. (2007). Three steps to understanding restorative environments as health resources. In C. Ward-Thompson & P. Travlou (Eds.), *Open space: People space 2* (pp. 163–80). Abingdon, UK: Routledge.

Hartig, T., & Staats, H. (2003). Guest editors' introduction: Restorative environments. *Journal of Environmental Psychology, 23*(2), 103–7.

Hartig, T., Mang, M., & Evans, G. W. (1991). Restorative effects of natural environment experiences. *Environment and Behavior, 23*(1), 3–26.

Hartig, T., Book, A., Garvill, J., Olsson, T., & Garling, T. (1996). Environmental influences on psychological restoration. *Scandinavian Journal of Psychology, 37*(4), 378–93.

Hartig, T., Evans, G. W., Jamner, L. D., Davis, D. S., & Gärling, T. (2003). Tracking restoration in natural and urban field settings. *Journal of Environmental Psychology, 23*(2), 109–23.

Hartig, T., Mitchell, R., de Vries, S., & Frumkin, H. (2014). Nature and health. *Annual Review of Public Health, 35*, 21.1–21.22.

Health Council of the Netherlands and Dutch Advisory Council for Research on Spatial Planning. (2004). *Nature and health: The influence of nature on social, psychological and physical well-being* (Vol. 2008). The Hague: Health Council of the Netherlands, RMNO.

Herzog, T., Black, A., Fountaine, K., & Knotts, D. (1997). Reflection and attentional recovery as distinctive benefits of restorative environments. *Journal of Environmental Psychology, 17*(2), 165–70.

Hine, R. (2007). *Ecotherapy: The green agenda for mental health.* London: MIND.

Hug, S.-M., Hartig, T., Hansmann, R., Seeland, K., & Hornung, R. (2009). Restorative qualities of indoor and outdoor exercise settings as predictors of exercise frequency. *Health & Place, 15*(4), 971–80.

Jiang, B., Chang, C., & Sullivan, W. C. (2014). A dose of nature: Tree cover, stress reduction, and gender differences. *Landscape and Urban Planning, 132*, 26–36.

Kaplan, R., & Kaplan, S. (1989). *The experience of nature: A psychological perspective.* Cambridge: Cambridge University Press.

Kaplan, S. (1995). The restorative benefits of nature: Toward an integrative framework. *Journal of Environmental Psychology,* (15), 169–82.

Kelly, V. C. (2012). A primer of affect psychology. In *The art of intimacy and the hidden challenge of shame* (pp. 158–91). Raleigh, NC: Tomkins Press.

Kuo, F. E. (2001). Coping with poverty: Impacts of environment and attention in the inner city. *Environment and Behavior, 33*(1), 5–34.

Kuo, F. E., & Sullivan, W. C. (2001). Aggression and violence in the inner city: Effects of environment via mental fatigue. *Environment and Behavior, 33*(4), 543–71.

Kuo, F. E., & Faber Taylor, A. (2004). A potential natural treatment for attention-deficit/hyperactivity disorder: Evidence from a national study. *American Journal of Public Health, 94*(9), 1580–6.

Laumann, K., Gärling, T., & Stormark, K. M. (2003). Selective attention and heart rate responses to natural and urban environments. *Journal of Environmental Psychology, 23*(2), 125–34.

Lewis, G., & Booth, M. (1994). Are cities bad for your mental health? *Psychological Medicine, 24*(4), 913–15.

Malone, K., & Tranter, P. (2003). Children's environmental learning and the use, design and management of schoolgrounds. *Children, Youth and Environments, 13*(2), 87–137.

Mayo Clinic Staff. (2013). Chronic stress puts your health at risk. Retrieved May 28, 2015, from www.mayoclinic.org/stress/art-20046037?pg=1

Mitchell, R. (2012). Is physical activity in natural environments better for mental health than physical activity in other environments? *Social Science & Medicine, online,* 1–5.

Mooney, P., & Nicell, P. (1992). The importance of exterior environment for Alzheimer residents: Effective care and risk management. *Healthcare Management Forum, 5*(2), 23–9.

Neisser, U. (1967). *Cognitive psychology.* Englewood Cliffs, NJ: Prentice Hall.

Nielsen, T. S., & Hansen, K. B. (2007). Do green areas affect health? Results from a Danish survey on the use of green areas and health indicators. *Health & Place, 13*(4), 839–50.

Orians, G. H. (1986). An ecological and evolutionary approach to landscape aesthetics. In E. Penning-Rowsel & D. Lowenthal (Eds.), *Landscape meanings and values* (pp. 3–22). London: Allen & Unwin.

Orians, G. H., & Heerwagen, J. (1992). Evolved responses to landscapes. In J. H. Barkow, L. Cosmides, & J. Tooby (Eds.), *The adapted mind: Evolutionary psychology and the generation of culture* (pp. 555–79). Oxford: Oxford University Press.

Orsega-Smith, E., & Mowen, A. (2004). The interaction of stress and park use on psycho-physiological health in older adults. *Journal of Leisure Research, 36*(2), 232–56.

Ottosson, J., & Grahn, P. (2005). A comparison of leisure time spent in a garden with leisure time spent indoors: On measures of restoration in residents in geriatric care. *Landscape Research, 30*(1), 23–55.

Park, B., Tsunetsugu, Y., & Kasetani, T. (2010). The physiological effects of Shinrin-yoku (taking in the forest atmosphere or forest bathing): Evidence from field experiments in 24 forests across Japan. *Environmental Health and Preventive Medicine, 15*, 18–26.

Parsons, R., Tassinary, L. G., Ulrich, R. S., Hebl, M. R., & Grossman-Alexander, M. (1998). The view from the road: Implications for stress recovery and immunization. *Journal of Environmental Psychology, 18*(2), 113–40.

Pressman, S. D., & Cohen, S. (2005). Does positive affect influence health? *Psychological Bulletin, 131*(6), 925–71.

Pretty, J., Peacock, J., Sellens, M., & Griffin, M. (2005). The mental and physical health outcomes of green exercise. *International Journal of Environmental Health Research, 15*(5), 319–37.

Pretty, J., Peacock, J., Hine, R., Sellens, M., South, N., & Griffin, M. (2007). Green exercise in the UK countryside: Effects on health and psychological well-being, and implications for policy and planning. *Journal of Environmental Planning and Management, 50*(2), 211–31.

Rohde, C., & Kendle, A. (1994). *Human well-being, natural landscapes and wildlife in urban areas: A review.* London: English Nature.

Scully, D., Kremer, J., Meade, M. M., Graham, R., & Dudgeon, K. (1998). Physical exercise and psychological well being: A critical review. *British Journal of Sports Medicine, 32*(2), 111–20.

Sousanis, J. (2011). World vehicle population tops 1 billion units. Retrieved May 28, 2015, from http://wardsauto.com/ar/world_vehicle_population_110815

Stigsdotter, U. K., Ekholm, O., Schipperijn, J., Toftager, M., Kamper-Jørgensen, F., & Randrup, T. B. (2010). Health promoting outdoor environments—associations between green space, and health, health-related quality of life and stress based on a Danish national representative survey. *Scandinavian Journal of Public Health, 38*(4), 411–17.

Sugiyama, T., Leslie, E., Giles-Corti, B., & Owen, N. (2008). Associations of neighbourhood greenness with physical and mental health: Do walking, social coherence and local social interaction explain the relationships? *Journal of Epidemiology and Community Health, 62*(5), e9–e9.

Taylor, A. F., Kuo, F. E., & Sullivan, W. C. (2002). Views of nature and self-discipline:

Evidence from inner city children. *Journal of Environmental Psychology, 22*(1–2), 49–63.

Thompson Coon, J., Boddy, K., Stein, K., Whear, R., Barton, J., & Depledge, M. H. (2011). Does participating in physical activity in outdoor natural environments have a greater effect on physical and mental wellbeing than physical activity indoors? A systematic review. *Environmental Science & Technology, 45*(5), 1761–72.

Tsunetsugu, Y., Park, B.-J., Ishii, H., Hirano, H., Kagawa, T., & Miyazaki, Y. (2007). Physiological effects of Shinrin-yoku (taking in the atmosphere of the forest) in an old-growth broadleaf forest in Yamagata prefecture, Japan. *Journal of Physiological Anthropology, 26*(2), 135–42.

Tzoulas, K., Korpela, K., Venn, S., Yli-Pelkonen, V., Kaźmierczak, A., Niemelä, J., & James, P. (2007). Promoting ecosystem and human health in urban areas using green infrastructure: A literature review. *Landscape and Urban Planning, 81*, 167–78.

Ulrich, R. S. (1979). Visual landscapes and psychological well-being. *Landscape Research, 4*, 17–23.

Ulrich, R. S. (1983). Aesthetic and affective response to natural environment. In I. Altman & J. F. Wohlwil (Eds.), *Human behavior and environment* (pp. 85–125). New York, NY: Plenum Press.

Ulrich, R. S. (1984). View through a window may influence recovery from surgery. *Science, 224*(4647), 420–1.

Ulrich, R. S. (1993). Biophilia, biophobia, and natural landscapes. In S. Kellert & E. O. Wilson (Eds.), *The biophilia hypothesis* (pp. 73–137). Washington, DC: Island Press.

Ulrich, R. S., Simons, R. F., Losito, B. D., Fiorito, E., Miles, M. A., & Zelson, M. (1991). Stress recovery during exposure to natural and urban environments. *Journal of Environmental Psychology, 11*(3), 201–30.

van-Dender, K., & Clever, M. (2013). *Recent trends in car usage in advanced economies—slower growth ahead*. Paris: International Transport Forum.

van den Berg, A. E., Koole, S. L., & van der Wulp, N. Y. (2003). Environmental preference and restoration: (How) are they related? *Journal of Environmental Psychology, 23*(2), 135–46.

van den Berg, A. E., Hartig, T., & Staats, H. (2007). Preference for nature in urbanized societies: Stress, restoration, and the pursuit of sustainability. *Journal of Social Issues, 63*(1), 79–96.

van den Berg, A. E., Maas, J., Verheij, R. A., & Groenewegen, P. P. (2010). Green space as a buffer between stressful life events and health. *Social Science & Medicine, 70*(8), 1203–10.

Vlastelica, M. (2008). Emotional stress as a trigger in sudden cardiac death. *Psychiatria Danubina, 20*(3), 411–14.

Whall, A., Black, M., & Groh, C. (1997). The effect of natural environments upon agitation and aggression in late stage dementia patients. *Journal of Healthcare Safety, Compliance & Infection Control, 3*(1), 31–5.

White, M. P., Alcock, I., Wheeler, B. W., & Depledge, M. H. (2013). Would you be happier living in a greener urban area? A fixed-effects analysis of panel data. *Psychological Science,* (April), 1–9.

Wohleb, E. S., Powell, N. D., Godbout, J. P., & Sheridan, J. F. (2013). Stress-induced recruitment of bone marrow-derived monocytes to the brain promotes anxiety-like behavior. *Journal of Neuroscience, 33*(34), online.

Zipes, D. P., & Wellens, H. J. J. (1998). Sudden cardiac death. *Circulation, 98*(21), 2334–51.

Chapter Eight

Social capital

The transition from the previous chapter on mental health to this chapter on social capital is a transition from the *intra*personal to the prominence *inter*personal relations play in an ecological model of health. These interpersonal relations affect one's social capital or "features of social organization such as networks, norms, and social trust that facilitate coordination and cooperation for mutual benefit" (Putnam, 1995, p. 67). What has been underappreciated in the extensive body of knowledge on the importance of interpersonal relations to health is how the physical environment can facilitate or hinder social capital, which has continually been shown to have a positive influence on both physical and mental health and well-being (e.g. Helliwell, 2003; Kim & Kawachi, 2007). Even more under-appreciated is how accessible GI can promote social capital. The relationship between GI and social capital is a microcosm of the reciprocity between the physical and social environments. People are part of the environment, our social actions influence the environment, and the environment in turn influences human health.

> the environment is a result of the constant interaction between natural and man [human]-made spatial forms, social processes, and relationships between individuals and groups. Many assume that the environment is everything outside of people. In fact, people do not exist or interact in limbo but in specific places which they in turn modify by their presence and activities. People are an essential part of the environment; they are as real as trees, rocks

or skyscrapers, and their interactions with each other and with places significantly influence their health and well-being.

(Lindheim & Syme, 1983, p. 337)

Previous chapters have presented how this reciprocity influences health directly (e.g. air, water) and indirectly through the behaviors that the environment supports (e.g. physical activity). Here we focus not on the human interaction with the physical environment per se but on the interactions between people that are facilitated or hindered by the environment. As inextricably linked as the natural environment is to the human-made built environment, so too are health-enhancing social relations reliant on the design of both of these types of environments. Our biophilic attachment to nature is evident in how social interactions are facilitated by the natural environment. We evolved encountering other organisms, including other humans, in the natural environment and formed social rules for civility and survival surrounded by nature. We proceed here with a very brief overview revealing that the connection between social capital, although not always called such, and health has been recognized for quite some time. This sets the stage for the many ways social capital has been measured and how GI supports the various domains of social capital. This is followed by a small but growing body of evidence focused specifically on how GI promotes the social capital critical for health.

The notion that the social environment can influence health is not new. Émile Durkheim's classic study of suicide in 1897 is often credited as the first to empirically test the relationship between the social environment and health (Durkheim, 1951).

Durkheim showed that even a behavior as individualistic and private as suicide could most usefully be understood in terms of the social setting within which it took place. He showed that people's integration into group life influenced the likelihood that they would be motivated to commit suicide. Those who were part of groups with weak social ties had higher rates of suicides than those living in groups with stronger ties. Further, even though people came and went from groups, the suicide rate for groups remained remarkably constant over time.

(Lindheim & Syme, 1983, p. 339)

There has been no shortage of criticism aimed at the limitations of Durkheim's study. Nonetheless, it established a social perspective to health focused on relationships. At the time, there was certainly no shortage of social explanations for disease—namely early public health theories that falsely attributed ill-health to the morally degenerate behaviors of those in low social classes—but these theories ignored the relationships that were taking place within the social environment that had a protective effect on health regardless of how saintly or ghastly one's morals. The reciprocity of the physical and social environments played no small part in this. One's relationships influenced the wretchedness of the physical environment one was exposed to.

Nearly a century after Durkheim, when the new public health was in the midst of readopting the ecological model, Lindheim and Syme (1983) argued for a greater public health focus on "meaningful social ties" (Cassel, 1976), recounting how social ties played a significant role in health outcomes ranging from tuberculosis to schizophrenia and alcoholism. Just over a decade later, Robert Putnam, building on the work of others (e.g. Coleman, 1988), popularized the idea of social capital. He believed that dwindling social capital was having an effect on a number of social ills including poor health. Like Durkheim, he operationalized social capital as membership in a group. Using evidence from a number of prospective studies done in a variety of countries and controlling for an array of individual hereditary and lifestyle factors, his newsworthy conclusion was that one's risk of dying the year after joining a group was cut in half and, by joining two groups, the risk was cut to a quarter (Putnam, 2001). Being part of something larger than oneself increased longevity. This is due to the effect of social relations, and resultant social capital, on the conditions and behaviors that increase the risk of premature mortality. Social capital undoubtedly influences the behaviors that contribute to many of the leading causes of death in high-income countries—for example, if one is less likely to perform physical activity—but there may be something about social capital itself, aside from the health behaviors that it instigates, that is essential for physical and mental health and well-being. Social capital may be tied to a sense of belonging or community attachment that supports health and well-being.

The significance of social capital to health is evident when it is compared to other behaviors and conditions more commonly associated with an increased risk of mortality. Robert Putnam has been credited with finding that "poor social capital is as bad as or worse than smoking, obesity, elevated blood pressure, or physical inactivity for human health" (Jackson, 2003, p. 193), and a more recent meta-analysis appears to support this claim. An analysis of 148 studies found a 50 percent increased likelihood of survival for those with stronger social ties consistent across age, sex, initial health status, and cause of death and that social relationships were as equally significant as other well-established risk factors of mortality (Holt-Lunstad, Smith, & Layton, 2010). It was concluded that "the influence of social relationships on the risk of death are comparable with well-established risk factors for mortality such as smoking and alcohol consumption and exceed the influence of other risk factors such as physical inactivity and obesity" (Holt-Lunstad et al., 2010, p. 20).

In their examination of the relationship between social capital and health, Cooper, Arber, Fee, and Ginn (1999) drew on the work of Coleman (1988) and Putnam (1995) to divide social capital into four categories. These categories elucidate the importance of GI on social health as they capture characteristics of both the social and physical environments.[1]

1. *Social resources*—Informal reciprocal support arrangements between neighbors, within and between friendship networks, and in specific "communities."
2. *Collective resources*—Level of civic activity as evidenced by community organizations, collective action and trust in institutions, and social cohesion.
3. *Economic resources*—Evidenced by opportunities for employment and the quality of environmental amenities.
4. *Cultural resources*—Quality of cultural amenities such as libraries, meeting places, and performance venues.

GI provides the physical environment necessary to support all four categories of social capital. First, interactions between people in GI can lead to an enhancement of the social networks and the *social resources* needed to bolster and sustain social capital (Figure 8.1). Green infrastructure acts as attractive meeting places for social activities that can strengthen and expand social networks. Second, GI can also act as a forum for increasing a community's *collective resources*. Using the example of the Green Gym movement in the UK, the civic activity of environmental restoration and conservation improves health through the physical demands of conservation activities but also through the social capital stemming from collective action. This brings with it other benefits to social capital. Thinking ecologically, the social capital fostered by GI might also be important to a feedback on the physical environment, protecting GI. Because environmental values are socially

Figure 8.1: Parks support a variety of social activities
Source: Elvert Barnes.

constructed (Dryzek, 1994; Proctor, 1998), heightening social capital in GI may be an important activity to form and foster shared environmental values that benefit the landscape and, in turn, human health. Improving environmental quality also improves the environment as an *economic* and *cultural resource*. Third, GI can support social capital as an economic resource not only by providing daily needs and the products of trade in subsistence communities but also in the larger scale exploitation of resources that provides goods and employment. Of course, the goal is that these practices are sustainable in that employment benefits are local and the exploitation of resources is sensitive to the capacity of the local environment. Green infrastructure is also an economic resource when considering the attractive aesthetic affect urban GI has on the perceived and actualized value of land and place. Fourth, some of this value is in GI as a *cultural resource*. Ranging from the local to the national, parks are viewed as defining features of neighborhoods and part of a national heritage. Parks contribute to creating a sense of place and community identity.

As is evident from the categories above, social capital is a broad term that captures a wide array of potential indicators of the social environment. This has fueled a justified criticism, and some suggested remedies, for advancing the variety of ways that social capital has been operationalized in health research (Almedom, 2005; Hawe & Shiell, 2000; Macinko & Starfield, 2001). The construct of *social cohesion* has been recommended to be used in place of social capital when considering GI and health because it is believed to be "more a characteristic of neighborhoods than of individuals (Baum, Ziersch, Zhang, & Osborne, 2009) and so more likely to be influenced by physical characteristics of the neighborhood, such as the availability and quality of greenspace and natural elements" (Hartig, Mitchell, de Vries, & Frumkin, 2014, p. 21.9). This is consistent with the Cooper et al. (1999) classification of social cohesion as a *collective resource*. Collective resources such as civic activity reflect the shared norms and values, friendly relationships, and the sense of belonging that contribute to social cohesion. While nature and health research may be heading in the direction of social cohesion, the studies cited below predate the call for a focus on social cohesion and therefore employ a variety of measures that aim to capture social capital but in slightly different ways. Given the current wide variety of terms used to describe the products and processes of the social environment (e.g. social cohesion, social support, social interactions, social relations, social ties, sense of community, etc.), the research cited below includes these terms as they were presented by the authors but with the understanding that they could be capturing slightly different aspects of social capital.

The ability of GI to support the various aspects of social capital as defined by Cooper et al. (1999) has been confirmed by a considerable amount of research examining the role of greenspaces in the social ecology of the urban poor in public housing. The results of this body of work confirms the original nineteenth-century beliefs and motivations for introducing parks in cities which were that greenspace was essential to the social and physical health of urban dwellers, particularly the most disadvantaged. Commenting on GI in the form of urban forestry, Kuo (2003, p. 153) states that disadvantaged urban neighborhoods are

> precisely the context where social ecosystem health is at greatest risk and where urban trees are least present. While poverty is not synonymous with alienation and risk of crime, too many poor urban neighborhoods are characterized by high levels of mistrust, isolation, graffiti, property crime, and

violent crime. It may be that the greatest benefits of urban forestry accrue to some of its historically most underserved constituencies.

Years of work by Kuo and her colleagues examining the role of greenspace on the social ecology of the urban poor support this claim.

A review of the work done by Kuo and her colleagues examining the link between GI in disadvantaged inner-city neighborhoods and social ecology found that GI was linked to a variety of positive social indicators including stronger ties among neighbors, greater sense of safety and adjustment, more supervision of children in outdoor spaces, healthier patterns of children's play, more use of neighborhood common spaces, fewer incivilities, fewer property crimes, and fewer violent crimes (Kuo, 2003). The communities that Kuo and her colleagues studied lacked communal facilities other than outdoor spaces and have reduced mobility due to poverty. Residents of these studies are therefore almost completely reliant on the outdoor communal spaces around their residences for social contact with neighbors. We proceed by looking at a number of individual studies by Kuo and her colleagues in this context and by others in different contexts. This evidence progresses from studies revealing a preference for green communal areas and the use of GI, to the contribution of GI to social capital, and finally to connecting the social capital brought about by GI to improved health.

First, a telling indicator of whether people are likely to commune in a particular space is their preference for design features that make spaces more attractive and therefore inviting. Green infrastructure in the form of trees is one such design feature. A photo simulation that compared people's preferences for communal areas in public housing complexes with varying levels of tree density found that there was a significant preference for communal areas with more trees (Kuo, Bacaicoa, & Sullivan, 1998). While not entirely shocking and hinting at a biophilic tendency, more telling and significant to the influence of GI on social capital is that one in three people stated they would use communal spaces more if trees were introduced. Now, preferences and intentions to use a communal space are a good starting point, but do these translate into the actual use of the space necessary for social interactions to occur? It appears so. Observations of these spaces revealed that greener areas attracted larger groups of more heterogeneous ages (Coley & Kuo, 1997). In a follow-up study, there was 83 percent more social activity in greenspaces versus more barren spaces with the pattern holding across age and gender (Sullivan, Kuo, & Depooter, 2004). Greenspaces were important in supporting social activity across age and gender, but they were particularly important for adults and females. Addressing the possibility that more social activity may simply be a product of a greater total number of people using greenspaces as compared to barren spaces, there was found to be proportionally more social than nonsocial activity in greenspaces when compared to more barren spaces.

The next step is then to determine if the use of greenspace has a positive effect on social capital. Among residents randomly assigned to 18 architecturally identical public housing facilities, there was found to be a positive association not only between increased levels of greenspace and use of common spaces but also between greenspace and a number of indicators of "neighborhood social ties" (i.e. amount of socializing in the building, familiarity with neighbors, sense of community) (Kuo, Sullivan, Levine-Coley, & Brunson, 1998). Furthermore, the relationship between greenery and neighborhood social ties was mediated by the use of the common spaces. In other words, more greenery led to more use, and this resulted in a positive effect on greater social ties in the neighborhood. The relationship between a greener environment and greater social capital also holds true among the elderly who are particularly prone to isolation. Examining the social ties of the elderly in public housing, it was found that those who lived in housing with greater exposure to green common areas had greater involvement with neighborly activities (e.g. talking with neighbors), reported stronger social relationships with friends and neighbors (e.g. greater familiarity with other residents), and had a stronger sense of local community (Kweon, Sullivan, & Wiley, 1998).

What is not clear is if the relationship between greener common areas and social capital would hold if there were other GI options for social interaction such as more public parks within walking distance of place of residence. Essentially what is being done by greening common areas is creating a park-like atmosphere, but this is no substitute for the public parks that are among the few public spaces available to all that can provide neutral gathering places for social interaction, recreation, and civic function (Dines, Cattell, Gesler, & Curtis, 2006). Parks are

an astonishing arena for forms of conviviality and collective pleasure. While parks and public space can also be danger zones at times, they are more likely to act as places where the rules of public life and citizenship are tested and formed. In this sense they are not just about improving the physical health and well-being of people as they go about their daily lives, but about creating more reciprocal forms of social life as well. There is no sustainable future without them.

(Warpole, 2007, p. 20)

We have known this for quite some time. As mentioned earlier, the perceived ability of urban parks to support social relations was a hallmark justification for their creation in the nineteenth century, and there was often an emphasis on how parks were particularly important to the most disadvantaged who had fewer options for escape from the brutal conditions of urban life at the time. Parks and park-like atmospheres are still much less common in the disadvantaged communities that arguably need them the most. While greening the immediate environment around one's home is good for social capital, it is not a substitute for increasing the availability of GI at larger scales, especially considering the multitude of other health co-benefits GI has been shown to provide. Until public GI in the form of parks and greenways pervades the urban environment, greening private property is a good first step to instilling the social capital green common spaces support, but it is not a substitute for connecting local GI to larger GI systems.

There appears to be something unique about greenspaces, as compared to other public spaces, that make them important in instigating social capital not only among the disadvantaged in urban neighborhoods but also among residents of "the burbs" (suburbs). Recall, it was the salubrious properties of the natural environment that helped spawn the rise of the burbs (Fishman, 1989). Although the burbs typically have more GI and everyday nature, suburban GI is predominantly on private property, which often comes at the expense of communal areas. Creating appropriate spaces for human interaction is essential for creating a sense of community as "community cannot form in the absence of communal space" (Duany, Plater-Zyberk, & Speck, 2000, p. 60). Comparing two types of suburban developments, a new urbanist neighborhood (that attempts to replicate many of the physical features of the urban environment thought to support community) to a traditional lower density suburban neighborhood, it was found that in both communities natural features were important to residents' *sense of community* (captured by the domains of community attachment, community identity, social interaction, and pedestrianism) (Kim & Kaplan, 2004). In fact, communal natural features (public greens, footpaths, children's playgrounds, and lakes and wetlands) were the only physical features that received consistently high ratings on sense of community among residents of both types of suburban communities. Furthermore, other design features of the new urbanist neighborhood (e.g. lot size, accessibility of sidewalks, street width) were significantly different from the traditional neighborhood in their perceived influence on social interaction. So, physical design makes a difference in perceived support for social interaction, but GI is important regardless of other design features that distinguish the two types of suburban communities.

The relationship between GI and social capital has also been found on greenway trails. Greenways have been found to be associated with increased pride in the community, more perceived opportunities for social interaction, and contributing to community identity (Shafer, Lee, & Turner, 2000). Taking greenways and other forms of GI together, a methodologically rigorous study revealed a relationship between the perceived levels of greenness (e.g. parks, paths with trees, other natural features) on not only local social interaction but also social cohesion (Sugiyama, Leslie, Giles-Corti, & Owen, 2008). We should recall that social cohesion is believed to better represent the contribution of the physical environment to social capital. One thing that has been lacking in the body of evidence cited up to this point has been the link between heightened social health and improved physical health. One such study remedies just this.

Green infrastructure appears to be important for facilitating social capital as measured in a wide variety of ways, and it is well established that social capital is a significant determinant of health. One study completes the chain of GI leading to heightened social capital leading to improved health. Looking first at greenspace and health, Maas, van Dillen, Verheij, and Groenewegen (2009) found that people with more greenspace in the immediate neighborhood environment (1km) had better self-perceived health, experienced fewer health complaints in the last 14 days, and had a lower self-rated propensity for psychiatric morbidity. Next, examining greenspace and social capital, they found that more greenspace in one's environment was associated with less feelings of loneliness (at 1km and 3km) and less of a perceived shortage of social support (at 1km). The relationship between greenspace and social support was strongest in the most urban communities and for youth, the elderly, and persons of low socioeconomic status, all believed to have lower levels of mobility. Taking these two pieces of information together, Maas et al. (2009) then explored the role of social capital in mediating the relationship between greenspace and health. In testing mediation, or the degree to which loneliness and social support intervened in the relationship between greenspace and health, it was found that loneliness within the immediate and larger environs around one's home partially mediated self-perceived health, health complaints in the last 14 days, and self-rated propensity for psychiatric morbidity. Shortage of social support in the immediate vicinity partly mediated the relationship between greenspace and health complaints in the last 14 days and more fully mediated the relation between greenspace and self-rated propensity for psychiatric morbidity.

Taking these findings together with findings by Kuo and her colleagues, and accepting the differences in the populations studied and the indicators of social capital, a chain of events leading from GI to social capital to health emerges. Kuo and her colleagues found that when nearby nature was present, social ties were heightened. Maas and her colleagues found that a lack of social support helped to explain the relationship between nearby greenspace and health. So, greenspace has a positive effect on social capital, and the social capital gained through accessible GI can positively affect physical and mental health.

Summary

GI plays an important role in promoting a healthy social ecology. In an ecological model of health, there is reciprocity between the social environment and the physical environment, and social capital is one indicator of the quality of this relationship. The relationship between the social and physical environments is evident in the role GI plays in promoting social capital. Green infrastructure also promotes social cohesion, which is believed to be a better indicator of GI on social health. The preference for green communal areas results in their increased use which leads to improvements in a number of indicators of social capital and mental and physical health. This is particularly true in disadvantaged communities that are further disadvantaged due to a typical absence of green communal spaces. Aside from the interactions derived from use, the presence of greenspaces is also a significant factor in instilling a sense of community identity and attachment that can contribute to well-being. While other, non-green public open spaces can also facilitate social capital, they can in no way rival the multitude of other health co-benefits of GI.

Note

1. This relationship between the two is apparent in a number of the conceptual models presented in Chapter 2 (e.g. the Health Map; Transformation via Balanced Exchanges (T-BE); Myers et al., 2013).

References

Almedom, A. M. (2005). Social capital and mental health: An interdisciplinary review of primary evidence. *Social Science & Medicine, 61*(6), 943–64.

Baum, F. E., Ziersch, A. M., Zhang, G., & Osborne, K. (2009). Do perceived neighbourhood cohesion and safety contribute to neighbourhood differences in health? *Health & Place, 15*(4), 925–34.

Cassel, J. (1976). The contribution of the social environment to host resistance. *American Journal of Epidemiology, 104*(2), 107–23.

Coleman, J. (1988). Social capital in the creation of human capital. *American Journal of Sociology, 94*, S95–S120.

Coley, R. L., & Kuo, F. E. (1997). Where does community grow? The social context created by nature in urban public housing. *Environment & Behavior, 29*(4), 468–94.

Cooper, H., Arber, S., Fee, L., & Ginn, J. (1999). *The influence of social support and social capital on health.* London: Health Education Authority.

Dines, N., Cattell, V., Gesler, W., & Curtis, S. (2006). *Public spaces, social relations and well-being in East London.* Bristol: Policy Press.

Dryzek, J. (1994). *Discursive democracy: Politics, policy, and political science.* New York, NY: Cambridge University Press.

Duany, A., Plater-Zyberk, E., & Speck, J. (2000). *Suburban nation.* New York, NY: North Point Press.

Durkheim, É. (1951). *Suicide: A study in sociology.* (G. Simpson, Ed.). Glencoe, IL: Free Press.

Fishman, R. (1989). *Bourgeois utopias: The rise and fall of suburbia.* New York, NY: Basic Books.

Hartig, T., Mitchell, R., de Vries, S., & Frumkin, H. (2014). Nature and health. *Annual Review of Public Health, 35*, 21.1–21.22.

Hawe, P., & Shiell, A. (2000). Social capital and health promotion: A review. *Social Science & Medicine, 51*(6), 871–85.

Helliwell, J. (2003). How's life? Combining individual and national variables to explain subjective well-being. *Economic Modelling, 20*(2), 331–60.

Holt-Lunstad, J., Smith, T. B., & Layton, J. B. (2010). Social relationships and mortality risk: A meta-analytic review. *PLoS Medicine, 7*(7), e1000316.

Jackson, L. (2003). The relationship of urban design to human health and condition. *Landscape and Urban Planning, 64*, 191–200.

Kim, D., & Kawachi, I. (2007). US state-level social capital and health-related quality of life: Multilevel evidence of main, mediating, and modifying effects. *Annals of Epidemiology, 17*(4), 258–69.

Kim, J., & Kaplan, R. (2004). Physical and psychological factors in sense of community: New urbanist Kentlands and nearby Orchard Village. *Environment and Behavior, 36*(3), 313–40.

Kuo, F. E. (2003). The role of arboriculture in a healthy social ecology. *Journal of Arboriculture, 29*(3), 148–55.

Kuo, F. E., Bacaicoa, M., & Sullivan, W. C. (1998). Transforming inner-city landscapes: Trees, sense of safety, and preference. *Environment and Behavior, 30*(1), 28–59.

Kuo, F. E., Sullivan, W. C., Levine-Coley, R., & Brunson, L. (1998). Fertile ground for community: Inner-city neighborhood common spaces. *American Journal of Community Psychology, 26*(6), 823–51.

Kweon, B.-S., Sullivan, W. C., & Wiley, A. R. (1998). Green common spaces and the social integration of inner-city older adults. *Environment and Behavior, 30*(6), 832–58.

Lindheim, R., & Syme, S. (1983). Environments, people, and health. *Annual Review of Public Health, 4*, 335–59.

Maas, J., van Dillen, S. M. E., Verheij, R. A., & Groenewegen, P. P. (2009). Social contacts as a possible mechanism behind the relation between green space and health. *Health & Place, 15*(2), 586–95.

Macinko, J., & Starfield, B. (2001). The utility of social capital in research on health determinants. *Milbank Quarterly, 79*(3), 387–427.

Myers, S. S., Gaffikin, L., Golden, C. D., Ostfeld, R. S., Redford, K. H., Ricketts, T. H., et al. (2013). Human health impacts of ecosystem alteration. *Proceedings of the National Academy of Sciences, 110*(47), 18753–60.

Proctor, J. D. (1998). The social construction of nature: Relativist accusations, pragmatist and critical realist responses. *Annals of the Association of American Geographers, 88*(3), 352–76.

Putnam, R. (1995). Bowling alone: America's declining social capital. *Journal of Democracy, 6*(1), 65–78.

Putnam, R. (2001). Social capital: Measurement and consequences. *Canadian Journal of Policy Research*, Spring, 41–51.

Shafer, C. S., Lee, B. K., & Turner, S. (2000). A tale of three greenway trails: User perceptions related to quality of life. *Landscape and Urban Planning, 49*(3–4), 163–78.

Sugiyama, T., Leslie, E., Giles-Corti, B., & Owen, N. (2008). Associations of neighbourhood greenness with physical and mental health: Do walking, social coherence and local social interaction explain the relationships? *Journal of Epidemiology and Community Health, 62*(5), e9.

Sullivan, W., Kuo, F., & Depooter, S. (2004). The fruit of urban nature: Vital neighborhood spaces. *Environment and Behavior, 36*(5), 678–700.

Warpole, K. (2007). The health of the people is the highest law: Public health, public policy and green space. In C. Ward Thompson & P. Travlou (Eds.), *Open space: People space* (pp. 11–22). Abingdon: Taylor & Francis.

PART III

CAUTIONS AND FUTURE DIRECTIONS

Chapter Nine

The threats to health posed by green infrastructure

An examination of the connections between GI and health would of course be incomplete without also examining some of the threats GI poses to health and well-being (Tomalak, Rossi, Ferrini, & Moro, 2011). The many forms of GI cannot be naively viewed as utopian oases that are free from risk. Focusing solely on the benefits of GI would be ignoring the necessary balance struck between life and death and the sometimes cruel and unpredictable ways that nature maintains this balance. Some of these threats are extremely rare and are mostly likely occur in wilderness areas. Few people live in these areas or will ever access them, but the presence of wilderness is essential to providing essential ecosystem services. Other more probable threats to health are posed when urban development encroaches on the countryside, when urban needs are met by exploiting regional or global natural resources, or when accessing the urban greenspaces that make city life healthier and more livable. Risks can range from a senescent tree limb falling on one's head to the more likely and manageable encounters with insects and exposure to allergens that can be addressed in many cases through awareness raising and through ecologically sensitive landscape design. Overall, the health benefits of GI conservation in most contexts far outweigh the risks, but the risks cannot be ignored.

Some risks stem from interactions with animals and insects. These risks include the rare yet chilling risk of being mauled or eaten by a predator, the risk of being injured due to the defensive actions of non-predators, and the risk of contracting zoonotic disease from interactions with animals that inhabit GI. A very small proportion of the world's population lives in areas where humans are not on the top of the food chain and therefore run the risk of becoming prey to carnivorous animals. This risk that was a part of human life for most of human history has now been nearly eliminated from the daily life of most humans due to the degradation of habitat and the more active extermination of the predators that could consider us food. In communities that still live with the daily stress of this threat, the solutions to reducing risk are the same as they always have been: alter the behaviors associated with an increased risk of encounters with predators or exterminate the predator. For the vast majority of humans that do not live with the daily threat of being eaten, potential encounters with predators are most likely to occur in or around the regional or national preserves to where predators have been isolated and where they play a vital role in a healthy ecosystem. A still rare but more likely risk stemming from interactions with non-domesticated animals in GI is not being mauled or eaten but rather being bitten or scratched. This can result in not only injury or death (a number of venomous snakes are prolific around my home in the Florida panhandle) but also the transmission of zoonotic disease. Discussed in the Zoonotic disease section of Chapter 5, disruption of habitat can increase the chances of transmission of zoonotic diseases predominantly through food and water resources being shared with animals and less likely through the aggressive or defensive actions of animals. While encounters with animals resulting in injury or death are very rare, there is a risk, and GI conservation must weigh these risks against potential benefits. What cannot be ignored is that some people will be put at disproportional risk for the benefit of the many. Those living in close proximity to the habitat of animals that could kill or harm will bear this risk for the benefit of the many who will reap the regional and global benefits of maintaining GI and the viability of ecosystem services. For example, those living in close proximity to national preserves where predators and their prey roam will live with the possibility of encounters with these animals while those in urban environments reap the benefits of air and water quality and climatic regulation from the presence of these large swaths of GI. Urban dwellers will put themselves at some level of measured risk when visiting the reserve, but doing so *could* bring with it local economic benefits to the people who live with daily threats. How these benefits are distributed is another story. This will be very context-specific and require very context-specific risk–reward analyses.

There is a much greater risk to health from encounters with the vectors that transmit infectious disease when humans access GI. Accessing all forms of GI could bring with it the increased risk of getting bitten or stung by disease-carrying arthropods such as ticks and mosquitoes, but, as was evident in the Vector-borne disease section of Chapter 5, GI that supports a healthy ecosystem actually reduces the risk of encounters with vectors. The ecology of these vectors is such that the risk of disease transmission increases when GI habitat is disrupted. With ticks, humans are more likely to become "accidental hosts" (Tomalak et al., 2011, p. 106) to disease when there is a lack of diversity of other vertebrates for ticks to feed on. Mosquitoes are more likely to transmit disease in deforested areas where their development is accelerated and where the habitat of the natural predators of mosquitoes (e.g. fish, reptiles, amphibians, and bats) is destroyed or degraded. Sterilizing greenspace by diminishing plant and animal diversity reduces the diluting effect that biodiversity has on decreasing the risk of vector-borne disease transmission to humans.[1] The biodiversity essential to ecosystem health is also good for humans and human health within the ecosystem.

Another threat GI poses to health stems not from creatures such as animals and insects but from those of the human variety. The masses of humanity moving into cities globally are insulated from some of the risks posed by GI, but this move also poses a new set of threats that can occur in GI. Segments of the population at higher risk of being assaulted (women, children, the elderly) by other humans (almost always men), experience greater barriers to realizing the health benefits of GI because of the real and perceived risks to personal safety. The empirical results from studies examining safety issues associated with GI are mixed, but it appears that the design and distribution of GI plays a significant part in reducing threats to safety. More greenspace in one's environment can be associated with increased feelings of social safety. Surprisingly, this is true for the aforementioned populations whose safety is typically thought to be in greatest jeopardy, women and the elderly, that actually feel safer in greener living environments (Maas et al., 2009). It is possible that the heightened social capital experienced in these environments contributed to increased feelings of safety, but this is not true of all types of greenspace. People in highly urbanized environments feel less safe in *closed* greenspaces (those that are densely forested and therefore reduce the ability to scan long distances). An analogy in the built environment might be a dark alley where potential threats are hidden and where those that deviate from social norms can stay out of view. This is significant because of the tension this creates between competing health benefits of GI. Negative perceptions of greenspace could influence usage patterns and therefore the potential benefits derived from accessing greenspace (Ward-Thompson et al., 2004). At the same time, "opening up" closed greenspace may degrade the habitat that reduces the risk of infectious disease. The unenviable task of balancing community needs with the context-specific ecology of GI is what is necessary to maximize the public health benefits of these spaces. As always, the health benefits of simple exposure to GI without accessing it are present regardless.

Another threat to health posed by GI is allergic reactions caused by a hyper-sensitivity to selected plants and insects. The most common allergen in nature is pollen from flowering plants, particularly grasses, with the most common reactions to these allergens being allergic rhinitis (sneezing, itching, runny nose), asthma (respiratory inflammation), and eczematous dermatitis (skin rash). While typically not fatal, these conditions can be debilitating and put a serious strain on the health care resources devoted to treating them. One might be inclined to think that the global migration to cities and subsequent separation from the natural environment might be leading to a reduction in the prevalence of allergies, but the opposite is actually true. "The prevalence of allergic diseases worldwide is rising dramatically in both developed and developing countries" (Pawankar, Canonica, Holgate, & Lockey, 2011, p. 1). A contributing factor to these trends could be the type of urban GI that an increasing proportion of the global population is being exposed to. Many non-native ornamental plants and trees installed into GI are highly allergenic and cause more harm than their naturally occurring counterparts (Lucadamo, 2011).[2] Choosing native vegetation over the non-native and purely ornamental will not eliminate risk, but it could reduce allergenic reactions and at the same time improve the ecological quality of GI (Figure 9.1).

Plants are the most common natural allergens, but allergic reactions (at times life-threatening) are also caused by the stinging insects (e.g. bees, wasps) that rely on plants as their food source and encounters with which are more likely in vegetated areas. Unlike the penetration of the skin by ticks and mosquitoes, stinging is a defensive mechanism that is not essential to insect development. This risk should not to be designed out of GI completely as these insects provide other important ecosystem services. The best solution to reducing risks in this case is therefore altering individual behaviors, namely identifying and avoiding the nesting areas of bees and wasps and ensuring that those with allergies to stings have the medication to counteract dangerous allergic reactions.

Another much less likely risk where individual behavior modification is the best solution to reducing risk is in the threat posed by the ingestion of toxic and poisonous mushrooms and plants. Similar to how awareness is raised of the

Figure 9.1: Mulberry tree in bloom

Source: F Delventhal.

Note: Mulberry trees introduced into the Arizona desert decades ago are responsible for widespread allergic reactions to their pollen. They, along with olive trees, are now illegal in some counties in Arizona.

risks posed by ingesting dangerous human-made products (e.g. lead-based paint), awareness of the threats from ingesting wild plants will likely go a long way in reducing risk. The lesson being, don't put things in your mouth unless you are sure they are safe.

Summary

Green infrastructure at scales ranging from the neighborhood park to the national reserves should not be naively viewed as utopian oases free from threats to health and safety. The natural environment can pose risks, but these risks can be greatly reduced by ecologically sensitive GI design that supports healthy ecosystems. Some risks are reduced by people changing their behavior and through being reintroduced to the rules of the natural environment. Green infrastructure is not an amenity where risks can be removed by eliminating it. The life-supporting and health-enhancing benefits of GI far outweigh the risks.

Notes

1. Appreciating ecologic complexity, and with some level of irony, the chemicals used to manage greenspace also pose risks to human health (Steingraber, 2002) both through direct contact with these often toxic chemicals but also through their effect on natural predators of vectors.
2. What is not clear is that non-native plants are native somewhere and if these plants in their native environment are less likely to produce allergic reactions either through increased human tolerance to them or by the reduced production of allergens in their native conditions.

References

Lucadamo, K. (2011, April 24). City's plan to plant million trees is pitting allergy sufferers versus aesthetics. *NY Daily News, Online.*

Maas, J., Spreeuwenberg, P., Van Winsum-Westra, M., Verheij, R. A., de Vries, S., & Groenewegen, P. P. (2009). Is green space in the living environment associated with people's feelings of social safety? *Environment and Planning A, 41*(7), 1763–77.

Pawankar, R., Canonica, G. W., Holgate, S. T., & Lockey, R. F. (2011). *WAO white book on allergy 2011–2012: Executive summary.* Tokyo: World Allergy Organization.

Steingraber, S. (2002). Exquisite communion: The body, landscape, and toxic exposures. In B. Johnson & K. Hill (Eds.), *Ecology and design: Frameworks for learning* (pp. 192–202). Washington, DC: Island Press.

Tomalak, M., Rossi, E., Ferrini, F., & Moro, P. (2011). Negative aspects and hazardous effects of forest environment on human health. In K. Nilsson, M. Sangster, C. Gallis, T. Hartig, S. de Vries, K. Seeland, & J. Schipperijn (Eds.), *Forests, trees and human health* (pp. 77–124). New York, NY: Springer Verlag.

Ward-Thompson, C., Aspinall, P., Bell, S., Findlay, C., Wherrett, J., & Travlou, P. (2004). *Open space and social inclusion: Local woodland use in central Scotland.* Edinburgh: Forestry Commission.

Chapter Ten

Concluding remarks

This book has summarized the diverse and complex ways that GI supports health. Some of the topics reviewed in this book were focused on ecological processes (e.g. the hydrological cycle) and the function of ecosystems (e.g. pollination). The study of these life-supporting processes and functions often does not take into account their vital importance to human health. Biophilic epidemiology research inserts the human into ecosystems to understand how the human influence on the environment (i.e. GI conservation) results in the ability of ecosystems to deliver the services that support health. These ecosystem services include those fundamental to survival and reliant on ecological and biophysical processes and the function of ecosystems (e.g. water, air, food), but these services also include those that can extend and enhance life (e.g. infectious disease regulation, physical activity, mental restoration). Most forms of GI simultaneously support myriad ecosystem services with numerous resultant co-benefits to health and well-being. For some, these co-benefits are obvious and tied to daily activities and meeting basic needs, but daily reminders of the connection between GI and satisfying basic needs are increasingly absent from the lives of the majority of the world's population now living in urban and urbanizing environments. Even in these largely artifact-filled environments, local, regional, and global GI is working in concert to meet basic needs while also supporting the health co-benefits of, and likely more pronounced need for, the ecosystem services of recreation, mental restoration, and social cohesion.

The natural environment has long been recognized as an important type of environment to health, but we are now in a period of unparalleled information feeding a constantly expanding litany of the health-supporting ecosystem services the natural environment provides to humans. Ecological models of health that include the natural environment correctly present it as fundamental to other spheres and constructs. Despite the encompassing and permeating influence of the natural environment on human health, it is the other spheres and constructs that continue to receive the greatest amount of attention and study. The now decades-old warnings of the diminishing state of the environment and ecological threats (Brundtland, 1987; Secretariat of the Convention on Biological Diversity, 2010) should be pushing public health to match the theoretical prominence of the natural environment with a commensurate effort in research and practice. Now, many decades after public health's acceptance of the ecological paradigm and a more accurate ecological conception of the critical interdependency between the natural environment and human health, what is overdue is a substantial increase in the intellectual and capital resources devoted to both understanding and conserving the GI that all humans depend on for their health and well-being. Conserving GI and creating an environment that supports health may help us narrow the differences in health that persist even after accounting for the more popular practice of isolating socioeconomic and demographic factors (Woolf & Aron, 2013).

An examination of GI for its human health benefits holds great potential to be the shared objective that encourages a greater consilience of the social, biological, political, health, and environmental sciences. Using the ebb and flow of collaboration between urban planning and public health as an example, collaboration occurred when the environment was deemed the culprit in a shared concern for improving public health (Sloane, 2006). It was a shared threat that united planning and public health, a change in the perception of threats that caused them to diverge, and it is a renewed appreciation for the environmental determinants of health that is uniting them again. Green infrastructure could be that critical contemporary issue that spans shared environmental and public health goals.

The transdisciplinary and interdisciplinary research revealing the connections between GI and physical and mental health is robust enough to support GI conservation as a health promotion strategy, but the body of evidence is not without its limitations. In a recent and systematic "review of reviews" of nature and health studies, Hartig, Mitchell, de Vries, and Frumkin (2014) note "strong agreement about the methodological state of the art" that has been dominated by observational studies. A strong foundation of research has been built, but the next step is to extend this research to more fully address gaps in our knowledge and the "noise" inherent in most study designs. The agreement among scholars of what needs to be done to advance nature and health research is summarized in the following list adapted from previous work identifying research gaps and needs (Bell, 2010; Hartig et al., 2014; Lachowycz & Jones, 2012; Sullivan, Frumkin, Jackson, & Chang, 2014).

1. *Greenspace typology and type of contact*—Not all GI is equal in its health-promoting potential. The characteristics of GI influence the potential health benefits derived from the presence of, and access and exposure to, GI. Creating a typology of GI that takes into account characteristics such as site-level design, distribution, system integration, and scale would allow nature and health research to be more readily comparable and cumulatively more productive. For example, at the scale of a neighborhood park, design characteristics can influence its attractiveness and use for physical activity and social interaction. Quality of GI adds predictive value to the health benefits of having larger quantities of GI close to home (van Dillen, de Vries, Groenewegen, & Spreeuwenberg, 2012). How integrated and connected the park is to other components of the GI system influences its accessibility and also the likelihood that its pervasiveness will increase population level exposure. The neighborhood park would also need to be considered as a component of a regional GI system. The presence of the GI system creates the landscape structure necessary for ecosystem functioning and ecosystem service provision.

   A typology created through a consensus of scholars would allow better communication and comparability of research and a critical starting point from which to continually refine the typology.[1] A generalizable typology would need to consider that some "green" infrastructure is not green at all but rather brown, such as in desert environments, and that the green and brown have a reciprocal relationship with blue infrastructure. The creation of a GI typology will also create more consistency between what is considered GI and what is termed the "natural environment." For example, if the natural environment includes all things not made by humans, is a monocrop of soybeans part of the natural environment? It is manipulated by humans, but not made by humans. The level of manipulation that may exclude a crop as part of the natural environment in some people's minds (e.g. genetic modification) may not be pertinent to excluding it as GI in nature and health research. Considering the health benefits of GI necessarily forces us to consider the health benefits of nature in all its forms. This includes distinctions between the built and natural environments. There are certainly biological and philosophical justifications for rejecting any differentiation between the human-made built and the natural environments, but this distinction could be considered more of a continuum than a hard line. The built environment is solely leaning towards the less natural end of the natural and built environment continuum. A typology may need to include levels of "naturalness" that include wilderness, the non-native tree on a city street, the urban garden, and industrial agriculture monocrop.

2. *Evidence from other disciplines*—It is my hope that the various sources of information pulled together for this book demonstrate the sweeping influence of GI on health and the variety of actors that will be necessary to study it in a way that accurately reflects its ecological complexity. It is admittedly trite to extol the need for interdisciplinary research, but if any field of study was worthy of a coming together of minds, it is GI and health. Using evidence from biology, atmospheric and earth sciences, ecology, and many other "hard" sciences to understand the fundamental processes that support life absolutely needs to be complemented by psychology, anthropology, political science, economics, urban and regional planning, and other social sciences to better understand the human relationship to the natural environment, how we relate to and value it, and the processes by which it can be protected (for our own health).

3. *Mechanisms and pathways*—The many theorized relationships and empirical associations between GI and health are being dissected to uncover physiological and social mechanisms or links in the pathway that starts with GI and ends with health and well-being. There is a healthy body of evidence revealing that access and exposure to GI is associated with various measures of health (e.g. mortality, psychiatric morbidity). This forms an important basis for digging a bit deeper. For example, there are decades of research confirming how social capital (or social cohesion) is important for health but much less on how the physical environment and GI can support more social capital. The next step is to understand how to get more people to use the GI that leads to greater social capital (and physical activity, reduced stress, etc.). Are the design features that make GI more attractive at odds with GI features that provide other essential ecosystem services? Increasing the amount of GI and creating a connected GI system may increase the size and viability of the system, but what are the political barriers to creating and conserving GI? All along the pathway that

leads from GI to health, there are needs for disciplinary expertise, but all this expertise needs to eventually lead to the larger issue of creating a landscape conducive to health. Case studies are often helpful in understanding how GI projects get done, the UK, Valencia, and Hamburg cases presented in Chapters 11, 12, and 13, respectively, acting as good examples.

4. *Significance of GI, duration of exposure, and thresholds effects*—The natural environment, ecosystems, and GI are the foundation of the ecological model of health. Despite this, the significance of GI to health is often overlooked in public health research and practice. A consistently more complete accounting of the numerous ecosystem services that GI provides is necessary to reveal the often hidden, yet sweeping, influence of the natural environment on health. Community GI assessments can reveal the permeating role of GI on health and how it can be conserved to better support community and global health. The duration of each ecosystem service benefit also needs to be better understood. Some benefits such as the contribution of GI to the hydrological cycle will be continuous, but others such as psychological restoration may require repeated and regular exposure to GI. Regional GI systems are necessary to provide continuous benefits, and local GI is necessary to achieve health benefits requiring regular exposure. Studying threshold effects will allow us to know if there is such a thing as too much GI. Is there a level of greenness after which there are diminishing returns? There is certainly a balance between reaping the health benefits of gray infrastructure development and the conservation of nature necessary for sustainable human habitation. Further examinations are needed to find the thresholds on the natural to built environment continuum where nature provides ecosystem services and the built environment can protect one against the risks to health associated with fickle nature. This is what Ebenezer Howard, and other early urban planners, was trying to achieve with his plans for Garden Cities over a century ago, but we are still far from realizing his dream of a healthy city.

5. *Varied effects on different population subgroups*—What is known to date about the relationship between GI and health is intimately tied to where the majority of studies have been conducted. We know a great deal about GI on a number of indicators of health in the Netherlands, mental restoration in Sweden, and social capital in Chicago inner-city neighborhoods, but other populations are not as well represented. This of course skews our understanding. For example, GI may not be as important to supporting physical activity in the Netherlands

where there is a wealth of other physical environment supports for physical activity and the cultural norm to use them, but GI may be vital to supporting physical activity in other locales where GI is the only safe space for recreation. While the health benefits of exposure and access will undoubtedly vary by population, the benefits of the presence of GI will not. All humans benefit from the ecosystem services of air and climate regulation that large-scale GI provides. It is the application of an ecosystem services approach to the practice of land-use planning and disturbances to the natural landscape (Niemelä et al., 2010) that brings to light the importance of GI to a healthy and sustainable future for all. All organisms, including humans, are reliant on GI to maintain life and health.

Note

1. I am the first to admit that I may be missing the mark here, and the complexity of creating a generalizable typology will require a meeting of the minds of those from various fields. I would be happy to contribute to any team effort to create this typology.

References

Bell, S. (2010). Challenges for research in landscape and health. In C. Ward Thompson, P. Aspinall, & S. Bell (Eds.), *Innovative approaches to researching landscape and health: Open space: People space 2* (pp. 257–78). London: Routledge.

Brundtland, G. (1987). *Our common future: Address to the World Commission on Environment and Development.* Tokyo.

Hartig, T., Mitchell, R., de Vries, S., & Frumkin, H. (2014). Nature and health. *Annual Review of Public Health, 35,* 21.1–21.22.

Lachowycz, K., & Jones, A. P. (2012). Towards a better understanding of the relationship between greenspace and health: Development of a theoretical framework. *Landscape and Urban Planning,* 8–15.

Niemelä, J., Saarela, S.-R., Söderman, T., Kopperoinen, L., Yli-Pelkonen, V., Väre, S., & Kotze, D. J. (2010). Using the ecosystem services approach for better planning and conservation of urban green spaces: A Finland case study. *Biodiversity and Conservation, 19*(11), 3225–43.

Secretariat of the Convention on Biological Diversity. (2010). *Global biodiversity outlook 3*. Montreal, Canada: Secretariat of the Convention on Biological Diversity.

Sloane, D. C. (2006). From congestion to sprawl. *Journal of the American Planning Association, 72*(1), 10–18.

Sullivan, W. C., Frumkin, H., Jackson, R. J., & Chang, C. Y. (2014). Gaia meets Asclepius: Creating healthy places. *Landscape and Urban Planning, 127*, 182–4.

van Dillen, S. M. E., de Vries, S., Groenewegen, P. P., & Spreeuwenberg, P. (2012). Greenspace in urban neighbourhoods and residents' health: Adding quality to quantity. *Journal of Epidemiology & Community Health, 66*(6), e8.

Woolf, S., & Aron, L. (2013). *US health in international perspective: Shorter lives, poorer health*. Washington, DC: The National Academies Press.

PART IV

SELECTED CASES OF THE HEALTH BENEFITS OF PLANNING AND IMPLEMENTING GREEN INFRASTRUCTURE

Chapter Eleven

Public health promotion in England's Community Forest Partnerships

Ian Mell

In 1990, the Countryside Commission for England established a network of Community Forest Partnerships (CFPs) across England. These 12 organizations were located in areas of social and environmental deprivation where industrial decline had a marked impact on quality of life (i.e. health) and the landscape (Kitchen, Marsden, & Milbourne, 2006). Over the course of their 25-year history, England's CFPs have acted as innovative land managers exploring the value of landscape enhancement and effective socio-environmental engagement to meet a range of health issues (Blackman & Thackray, 2007; Coles & Bussey, 2000). As previous chapters have explored, the application of green infrastructure (GI) to improve health takes many forms. The CFPs, throughout their history, have aimed to deliver multifunctional landscape resources that address the interactivity of climate change, biodiversity, water management, and health collectively.

Located in close proximity to a number of England's post-industrial cities including Manchester, Liverpool, and Newcastle, England's CFPs were endowed with a remit of revitalizing the socioeconomic and environmental value of depressed areas (Forest Research, 2010). However, unlike other spatially defined initiatives—for instance National Parks—CFPs were not defined by a specific physical boundary. Alternatively, they were developed to work within more flexible spatial boundaries that reflect the socioeconomic disparities loosely aligned to local government boundaries. As a consequence, CFPs cannot be defined as "forests" in the traditional sense. Rather, they should be considered as land managers, social enterprise innovators, and advocates of environmental improvement. Figure 11.1 presents the locations of the 12 original CFPs. Figure 11.2 presents the eight partnerships that remain post-local government funding changes in 2008–09.[1]

In terms of their management and focus, CFPs are charitable organizations working within the environmental sector. They work with a range of academic, government, and development partners to improve the physical and social structure of post-industrial landscapes. The "forests" are made up of non-contiguous land

Figure 11.1: Location of original Community Forest Partnerships in England
Source: Adapted from England's Community Forests.

units, some of which are owned by local government. Others are developed and managed under land agreements with private owners whose land is located within the spatially defined community forest boundary. The "partnerships" themselves are responsible for the development of investment and management programs that utilize GI to meet a number of socioeconomic and environmental issues. Each partnership should be considered as a collaborative advocacy organization that works with other agencies to foster change in land use within the more abstract community forest boundary (England's Community Forests, 2004; Mell, 2011).

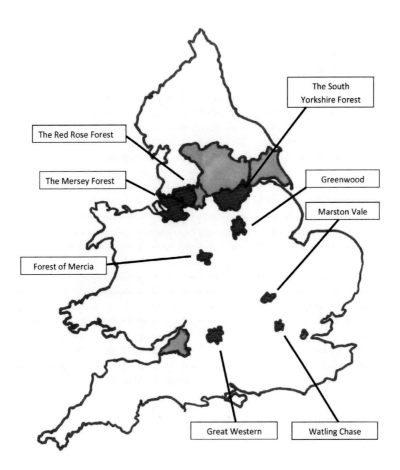

Figure 11.2: Location of existing Community Forest Partnerships in England
Source: Adapted from England's Community Forests.

With regards to the promotion of health initiatives, each partnership has and continues to maintain collaborative networks with health practitioners (e.g. National Health Service (NHS), Primary Care Trusts (PCTs), and local communities) to promote a more inclusive functionality of local landscape resources.

Over the course of their existence, each CFP has developed successful relationships with local partners and communities leading to visible improvements in the level of awareness of health inequalities. They have also influenced the subsequent reactions in policy and practice at a local authority level (Blackman & Thackray, 2007). The focus of these projects ranges from engagement with individuals to improve their awareness and confidence in using GI to the more extensive promotion of programs aimed at improving mental and physical health (Mell, 2007, 2011). The outcome of this process has varied across each partnership; however, the breadth of project work undertaken by each CFP has increased as the NHS and PCT partnerships have engaged more directly with academic and practitioner evidence demonstrating the positive impact the natural environment has on health (Maas, Verheij, Groenewegen, de Vries, & Spreeuwenberg, 2006; Pretty, Peacock, Sellens, & Griffin, 2005; Ulrich, 1984).

The successful formulation and delivery of GI health initiatives are evident across the many CFPs. Furthermore, a spatially cohesive group of CFPs in the north of England have managed to successfully diversify, expand, and deliver value across a number of depressed locations (McDermott & Schreckenberg, 2009). The M62 highway corridor from Liverpool to Leeds is home to eight of the ten most deprived communities[2] in England; communities that illustrate visible signs of extreme health inequality between communities and individuals (TEP, 2005). It is also the location for three of the most successful CFPs: The Mersey Forest, Red Rose Forest, and the South Yorkshire/White Rose CFP. The following sections present the value of GI in addressing health issues at a number of scales using examples from each of these partnerships, as well as from the North East Community Forest which in the mid-2000s helped to shape the development of GI/health projects across North East England.[3] The links between the underuse and undervaluation of landscape resources and deprivation/health inequality are proposed as central elements of community forestry activities in England. In response to these issues, each CFP has attempted to deliver projects that make quantifiably positive impacts on both the local environment (i.e. physical enhancement) and the health (physical and mental well-being) of local populations.

England's Community Forest Partnerships

The function of each CFP (Table 11.1) is to respond to the socioeconomic dislocation of people and the landscape in former industrial locations (Blackman & Thackray, 2007; Mell, 2011). As a result, their spatial distribution locates them in close proximity to many former industrial cities meaning that they are situated in areas with high population density, with many classified as having high levels of socioeconomic deprivation. Each CFP works extensively with a range of local government authorities and public–private organizations to facilitate landscape improvements and more efficient landscape management. This is particularly important in former industrial areas where there has been a historical under-funding for investment in landscape enhancement.

Table 11.1: England's Community Forest Partnerships

Forest and location	Size	Landscape, access, and recreation context
Red Rose Forest Covers a large area of Greater Manchester	292 miles^2	Covering the broad, flat basin of the Mersey floodplain, the forest area is predominantly urban. It is a largely flat, densely populated area with a multiplicity of development types and a comprehensive transport infrastructure.
The Great North Forest★ Located across south Tyne and Wear and northeast Durham	96 miles^2	The majority of the forest area is under some form of agricultural management, mainly arable cultivation. The legacy of coal mining is also apparent. At its inner boundary the forest connects with the green corridors and open spaces that permeate the urban fabric but is less well defined at its outer edge where it merges with the surrounding wider rural landscapes.
The Tees Forest★ Located in the Tees Valley surrounding the urban areas of Darlington, Middlesbrough and Stockton-on-Tees and stretching to the coastal settlements of Redcar, Hartlepool and Loftus	135 miles^2	The majority of the forest area is an intensively farmed countryside interspersed with modern industry, transport routes, and housing. The local landscape affords attractive views of the North York Moors but also intrusive features like power stations and transmission lines and abrupt edges to housing developments. There is an extensive urban fringe where developments and utilities infrastructure intermingle with agriculture and relics of high quality landscape.

Forest and location	Size	Landscape, access, and recreation context
The Mersey Forest Takes in the nine local authorities of Merseyside as well as North Cheshire	420 miles^2	The forest area has a clear functional identity based around the Mersey Estuary at its core and the large-scale industrial conurbation that has grown up around it. The urban fringe landscape reflects the area's industrial past, with a significant amount of derelict land some of which is unsuitable for house building or un-remediated.

Source: Adapted from Land Use Consultants with SQW Ltd. (2005).

Note: *Although both The Great North Forest and The Tees Forest ran independent community forest investment programs, they were structurally managed jointly as part of North East Community Forest (NECF) Partnership from the NECF Head Offices in Whickham and then Annfield Plain, England.

Red Rose Forest

The Red Rose Forest (Figure 11.3) is actively involved in land management and GI development in central and western Greater Manchester. The area served by the Red Rose Forest has a population of approximately 1.75 million and covers an area of approximately 292 square miles. They have worked extensively with local communities, businesses, and other public–private partners to develop multipurpose landscapes that help to create a better quality of life and place (Figure 11.4). The Red Rose Forest has eight strategic investment/management priorities, within which the promotion of improved physical and mental health is central. The Red Rose Forest has made strategic investments in walking, cycling, and biodiversity infrastructure across each of the local government authority areas that work within Greater Manchester (Salford, Wigan, Bury, Bolton, Trafford, and Central Manchester) to address various climatic, economic, and social aspects of health improvements.

North East Community Forest Partnership (The Great North Forest and The Tees Forest)

The North East Community Forest (NECF) was a regional partnership comprised of The Great North Forest (TGNF) and The Tees Forest (TTF) (Figure 11.5).[4]

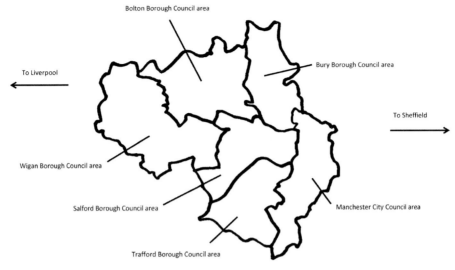

Figure 11.3: Red Rose Forest

Source: Adapted from England's Community Forests.

Figure 11.4: Red Rose Forest green roof activity

Source: Red Rose Forest Community Partnership.

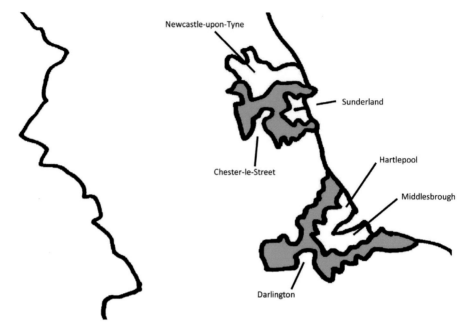

Figure 11.5: North East Community Forest
Source: Adapted from England's Community Forests.

TGNF was located in the urban fringe of Newcastle and South Tyneside, while TTF served the Teeside region including Middlesbrough, Stockton, Hartlepool, and Darlington. TGNF covered an area of 96 square miles while TTF had a spatial area of 135 square miles. The area that was served by the larger NECF has a population of approximately 1.41 million. The NECF frequently worked with local authorities, private landowners, communities, and developers to improve the quality of the built environment through investments in GI. They were one of the first organizations (along with The Mersey Forest) to develop GI guidance on the development and management of forestry and urban greening (Davies, Macfarlane, McGloin, & Roe, 2006). They were also one of the first agencies to work directly with NHS and PCT partners to promote the health benefits of engagement with GI resources. In 2008 the NECF ceased trading with its public engagement and tree planting programs taken over by Groundwork North East (Mell, 2011). The successful GI investments delivered by NECF are still viewed as laying the foundation for more recent projects in North East England (Mell, 2011).

The Mersey Forest

The Mersey Forest (Figure 11.6) is spatially the largest of the designated community forests in England. The organization has a working boundary that covers 420 square miles of Merseyside and North West Cheshire and works with each of the seven local authorities in this geographical area (Cheshire West and Chester, Halton, Knowsley, Liverpool, Sefton, St. Helens, and Warrington). The area served by The Mersey Forest has a population of approximately 1.8 million. Over their history, The Mersey Forest Partnership has planted over eight million trees. They have also worked extensively with local communities, developers, and local government to redevelop derelict and undervalued landscapes into multifunctional

Figure 11.6: The Mersey Forest

Source: Adapted from England's Community Forests.

and valued public assets. The Mersey Forest played a central role in the development and implementation of GI policy across the community forest area, and more widely in the North West of England. One of the key delivery objectives of The Mersey Forest Plan (Mersey Forest, 2013b) states their aim to improve health by engaging people with GI, as well as planting trees in urban and urban-fringe areas to create more attractive, walkable, and multifunctional neighborhoods.

The following sections discuss a number of GI investments by the CFPs that have either been health focused or used to facilitate health projects and initiatives.

Natural Choices for Health and Wellbeing

In 2011, the Natural Choices for Health and Wellbeing program was initiated by The Mersey Forest as part of the Decade of Health and Wellbeing. The project evolved from the Liverpool Green Infrastructure Strategy (Green Infrastructure North West, 2010), which identified where health inequalities were located across Liverpool. To facilitate the project, public health transition funding was allocated through the PCTs to help local community groups and charities invest in GI. The project encouraged participation with the "Five Ways of Wellbeing" (*connect, be active, take notice, keep learning,* and *give*) to facilitate a behavioral change in participants. The Mersey Forest assisted in the administration of the project, utilizing their networks of partners, community groups, and volunteers to distribute £296,000 of funding allocated to 38 local projects (Mersey Forest, 2013a).

An evaluation of the projects undertaken by the University of Essex for The Mersey Forest/NHS illustrated an 18 percent rise in self-assessed well-being amongst participants compared to 10 percent on similar projects not focused on GI. It also highlighted that participants believed that their well-being had improved, showing a marked increase in individual "above average" well-being compared to the average for Liverpool. Moreover, due to the changes to the physical landscape, 85 percent of people who engaged with the Natural Choices program took part in some form of physical exercise, of which 19 percent reported a visible improvement in health. Furthermore, over 80 percent of the projects saw a longer term increase in physical activity as one of their main project outcomes (Mersey Forest, 2013a).

The outcome of the project saw an increase in the number of small-scale community projects across Merseyside and North Cheshire bidding for localized funding. This manifested in the development of a number of community allotments, gardens, and orchards. The project also funded a range of wildlife areas and health-awareness raising projects. The long-term impact of the project has been, first, the development of more physical spaces to promote engagement with the environment, and, second, an increased capacity of local people willing to improve their own health through working with Natural Choices projects (England's Community Forests & Forestry Commission, 2012). From a health perspective, project coordinators noted the improved well-being (mental, physical, and social) of participants who took part in the projects, as well as an increase in longevity through engagement in physical exercise.

Running Rings

Developed to coincide with the positive social attitudes surrounding the London 2012 Olympic Games and the Inspire 2012 program, The Mersey Forest Running Rings project was established to encourage people to use their local community forest to improve their long-term health. The project made extensive use of local GI investments at Mab Lane (Liverpool) and Bold Forest Park (St. Helens). It also made use of Mersey Forest funding and managed greenways and woodlands to plan a range of accessible routes allowing different user groups to participate in group activities. The Mersey Forest also worked in collaboration with Warrington Borough Council and a local running club, Spectrum Striders, to target as wide a range of people as possible to participate, as an addition to their existing Green Gym (see Chapter 2 and upcoming Green Gyms and health walks section) activities.

Forest Schools

Forest Schools have become an increasingly popular mechanism for the CFPs to address the issues of Nature Deficit Disorder discussed by Louv (2005), and encourage a long-term appreciation of outdoor environments. Using the Youth Physical Activity Promotion Model developed by Welk (1999), The Mersey Forest, with Liverpool John Moores University, assessed the value of Forest Schools on the health and well-being of children engaged in the program's classes between March and July 2009 (Ridgers & Sayers, 2010). It was found that exposure to outdoor environments, which utilized a mixture of planned and adventure learning, led to greater engagement with the landscape in both children and parents (Figure 11.7).

Participation with the program was also seen to encourage changes in the behavior of children who took part by allowing them more freedom to learn leading to more frequent and longer term use of outdoor spaces. Further analysis of the Forest Schools program found a 7.8 percent increase in pro-social interactions, a longer engagement with play and outdoor activities, and an improved retention of the key health, education, and play message presented during the activities (Ridgers & Sayers, 2010).

Green Gyms and health walks

Green Gym programs were run by NECF with the assistance of specialist health officers (NHS and PCT) in conjunction with the British Trust for Conservation Volunteers (BTCV) on a number of community forest sites. The programs found that 90 percent of participants in outdoor activity saw an improvement in physical and mental health during their participation. These successes were supported by further reports of improved physical and mental well-being associated with continued outdoor activity (Figure 11.8) (Forest Research, n.d.).

Figure 11.7: Children in Forest School
Source: Gemma Jerome.

Figure 11.8: Outdoor exercise achieved through conservation work

Source: Gemma Jerome.

A series of evaluations of NECF Green Gym programs highlighted the value of accessible (and multifunctional) GI in promoting and subsequently facilitating engagement with formal and informal exercise programs (Forest Research, 2010). When compared to the national evaluation (Yerrell, 2008), the NECF evaluations showed comparable positive evidence of the value of Green Gym activities. In the BTCV study, participants were asked to reflect on the benefits of Green Gym in the community. In response participants "agreed" or "strongly agreed" with the statements that Green Gym programs had a positive impact on "Health and confidence" (99 percent), "Skills and training" (94 percent), and "Contribution to the environment" (92 percent).[5]

Furthermore, participants in Green Gym programs were more likely to have an increased longevity as compared to those that followed traditional gym-based regimes (Brown and Grant, 2005). One such program successfully engaged the Orthodox (Haredi) Jewish community in Gateshead, South Tyneside. As a community that is often considered "hard to reach," the Green Gym program aimed to first engage and then raise awareness of the potential health benefits of exercise (indoors and outdoors) (Gateshead Council, 2005). The programs also aimed to break down existing resistance within the Jewish community to exercising in public. Although the funding for this program limited its long-term viability, NECF staff reported that some of the existing social barriers to accessing health activities started to break down because of the program.

Moving from an urban to an urban-fringe location to the west of Gateshead, but within The Great North Forest area, Chopwell Wood was promoted by a consortium of local authorities, PCTs, and environmental agencies (i.e. Forestry Commission) as a Pilot Health Project. Woodland officers worked in partnership with stakeholders to facilitate access to GI for the Gateshead Exercise Referral Programme (Forest Research, 2006) and included the provision of Green Gym, health walks, cycling, and Tai Chi classes. The activities were also promoted to the general public but were specifically aimed at hard-to-reach groups including the Orthodox Jewish community in the Saltwell Park area of Gateshead. The outcomes of the project saw a measurable improvement in the health of participants as well as increased participation in both formal and informal physical activity (O'Brien & Snowdon, 2007; Snowdon, 2005).

A further successful program was the "Walking for Health" initiative which promoted inclusive and informal forms of exercise to aid those people who felt excluded (self or peer) from formal exercise (O'Brien, 2005; Snowdon, 2005).

The walks initially offered by health practitioners and NECF officers, and now by local authority and health officers, provided a guided experience with other "like-minded people" allowing all demographic groups to participate in low-stress exercise (O'Brien & Snowdon, 2007). Working in partnership with the Ramblers Association and Macmillan Cancer Support enabled the projects to develop within friendly and knowledgeable environments. Several evaluations of the project reported positive feedback including 90.9 percent of people referred to the project completing the full 13 weeks of activities. An additional success was that a further 128 people self-referred and participated in walking, cycling, and Tai Chi activities after seeing publicity about the project (O'Brien & Snowdon, 2007).

A number of similar programs were also developed by NECF across The Great North Forest and The Tees Forest areas. In Hetton Lyons, weekly walking and cycling clubs were organized to make better use of Hetton Lyons Country Park in order to meet social and health disparities in County Durham (Mell, 2006). While in Teeside, The Tees Forest promoted "health walks" in Albert Park (Middlesbrough) and South Park (Darlington) to facilitate engagement with health issues in a range of groups including young adults, single parents, and older people—people who had been identified by the NHS as being at risk of poor health (Natural England, 2011).

Healthy Schools

Chopwell Wood was also the location of a Healthy Schools program which engaged four local schools in the Derwentside area. Each school made four visits to the woods to undertake physical activity, learn about nutrition, healthy eating, and the value of outdoor activity in well-being (i.e. stress reduction). The project evaluation highlighted that, post-visit, 87 percent of children felt that the woodland was a place where they could be healthy and where they could engage in fun health activities (compared to 74 percent pre-visit). This subsequently led to a greater engagement and use of the woodland by local children and families who continued to engage in outdoor activities (Forest Research, 2006). Further evaluation of parent and staff responses illustrated a strongly held belief that the program had a positive influence on children/young people. It noted that repeated engagement with the program led to:

1. Increased awareness of nutrition and healthy lifestyle.

2. Increased knowledge, awareness, and appreciation of nature and the environment.
3. Increased levels of physical activity.
4. High level of recall amongst young people due to their engagement in the woodland-based activities, which facilitated learning.

The outcomes of the project highlight the role played by awareness and engagement with GI, in this case woodlands, to enable children/young people to participate in formal health-based activities (O'Brien & Snowdon, 2007; Snowdon, 2005).

Green Streets projects

Across The Mersey Forest and Red Rose Forest, both partnerships have invested heavily in tree planting activities. The Red Rose Forest invested in landmark street trees in the Oxford Road area of Manchester to address climate change and environmental aesthetic issues. The outcomes of this investment reported by local residents and businesses were that the trees made the area greener and therefore increased liveability, which has additional health and well-being benefits for the local population (Mell, Henneberry, Hehl-Lange, & Keskin, 2013; South Yorkshire Forest Partnership & Sheffield City Council, 2012).[6] In addition to street tree planting, the Red Rose Forest also facilitated the development of community gardens, allotments, and orchards in areas of health inequality to allow local people to work with the environment to pursue their own health initiatives (Red Rose Forest, 2014). The outcomes of such projects were reported to have aided improvement in personal and communal quality of life, as well as promoting social cohesion. These projects used established spaces to engage local people with these activities; furthermore, they made use of "Meanwhile" spaces to foster short–medium term reassessments of the value of the landscape (Red Rose Forest, 2013).

In Merseyside and North Cheshire, The Mersey Forest used £0.5 million from the Local Sustainable Transport Fund (LSTF) and the UK Department of Transport (DoT) to plant over 1,000 trees (from 2007 onwards). Trees were planted to increase awareness of the use of alternative (non-motorized) forms of transport by making visible improvements to the physical environment. Through a series of tree planting activities, The Mersey Forest looked to promote the idea that better quality cycling and walking infrastructure would lead to increased use and, by association, better health and well-being. Greening of commuter and recreation routes also helped to deliver climate change benefits (pollution and rainfall capture, retention and release), which had a direct impact on the quality of the local environment and health (Mersey Forest, 2013b).

England's Community Forests and health: reflection

Throughout the 1980s and 1990s, a number of policy responses were proposed by the UK government to revalorize the role of England's landscapes, a number of which reflected the growing health inequalities identified around former industrial cities (DTLR, 2002; Urban Task Force, 1999). However, the perceived dislocation between specific places and society revealed clear variations in the success rates of associated health initiatives. One initiative that has continually grown over that period has been the investment in GI through community forestry in England. England's CFPs have acted as cost-effective innovators that work with a broad range of health and government agencies to deliver better quality environments which address health issues (qualitatively and quantitatively) within communities of need (Department of Health, 2004; Mell, 2011).

One significant aspect of this process has been the effective engagement by England's CFPs, especially The Mersey Forest, with the economic case for linking GI to health. Reflecting on cost–benefit analysis of investment in GI versus primary–tertiary health care costs, a number of researchers (e.g. Pretty et al., 2007, 2005) proposed a strong quantitative economic rationalization for investing in GI. As evidence has become progressively more refined, GI research has identified the economic savings that increased physical activity could have for the UK NHS (Mell, 2007). An indicative figure of approximately £2 billion per year is the cost of inactivity: 2–3 percent of the NHS annual budget (Department of Health, 2004); obesity costs a further £1 billion in direct, and £3 billion in indirect, costs, while diabetes has a £1.3 billion cost to the UK economy (Natural Economy Northwest, 2008). The cost to the UK economy of mental health problems (mild and acute) has been estimated to be approximately £23 billion per annum. Considering one in six adults have been diagnosed with some form of mental health issue in the UK, developing strategies to minimize this impact has been a priority for the NHS (Layard, Clark, Knapp, & Mayraz, 2007; Liverpool City-Region Health is Wealth Commission, 2008). Furthermore, the UK also has one of the highest rates of childhood asthma (15 percent) in the world, which is particularly prominent in areas of deprivation and lower socioeconomic standing. Effective redevelopment of derelict/disused landscapes to improve the quality of urban and urban-fringe environments through investment in GI can thus be viewed as reducing the exposure to pollutants and incidence of asthma attacks. As a result, locations that invest in GI (e.g. street trees) are more effective in addressing health effects directly (Lovasi, Quinn, Neckerman, Perzanowski, & Rundle, 2008).

Responding to the economic perspectives of the NHS, CFPs were created to facilitate a behavioral change in the valuation, management, and subsequent use of urban-fringe areas with a view to establishing a healthier population. Aligned with this process has been the need to create a socioeconomic business case to attract finance to "investment black-spots" to enable funding to be drawn down and facilitate locally significant investments in GI. Over their 25-year history, CFPs have, in part, successfully created an environment where investment has been delivered to address social exclusion, underemployment, and health inequality based on a re-establishment of more positive human–environmental interactions (Blackman & Thackray, 2007; Louv, 2005).

The value to health of community forestry in England therefore draws on the versatility that investment in GI provides for CFPs. As investors in landscape enhancement, they are well placed to integrate myriad policy mandates into deliverable enhancements of the physical environment. In terms of addressing health problems, CFPs have utilized their position as external contractors and/or collaborators, including local planning authorities, to actively engage more directly with those communities identified as showing health inequality (Blackman & Thackray, 2007; Coles & Bussey, 2000). Improvements to the physical structure of the landscape, coupled with greater accessibility and a growing socio-environmental awareness of local populations, has created an arena where health issues can be addressed through project work. As the examples presented in this chapter illustrate, the variety of projects developed by CFPs has enabled them to modify existing approaches to land management by engaging directly with the major health issues of communities. Their engagement with existing (i.e. Chopwell Wood) and new (i.e. Oxford Road street trees) GI resources has enabled each partnership to target the specific health needs of a given location. Therefore, although GI cannot be considered as an *ex-ante* "one size fits all" solution to heath issues, it does broaden the scope of the discussion about what form of investment (physical, behavioral, and financial) is possible.

Engaging hard-to-reach groups and effectively managing referral participants has been a further success of health-oriented projects run by CFPs. This illustrates the ability of CFPs to narrow the gap between the identification of health problems and an active engagement in activities that make best use of GI at a local community scale. Furthermore, the structure of their partnership arrangements as key collaborative advocacy/environmental agencies provide each CFP with a diverse range of host partner agencies to work with in order to target, engage, and facilitate better health.

While the longevity of some health initiatives may be limited by the funding available for project work, investment in GI provides a physical legacy that continues to be seen as engaging local communities *ex-post*. The projects discussed in this chapter highlight a number of opportunities afforded by GI to promote improvements in health through participation in community forestry activities. They also suggest that building local capacity through structured engagement has achieved a longer term relationship between people and the environment. This should lead to a more sustained improvement in health.

Summary

Reflecting on the added value achieved by CFPs, it is unclear whether any other advocacy agency has played such a pivotal role in re-establishing the value of nature to health in England. While a cyclical re-engagement with the Garden Cities ideals of the Town and Country Planning Association could be considered a biannual event in the UK (Town and Country Planning Association, 2012), the longevity of community forestry in England illustrates the values of consistency, application, and the innate ability of each partnership to diversify and address the most prominent issues of the day (Blackman & Thackray, 2007; Mell, 2011). Community forestry in England has not attempted to simply repackage environmentalism through tree planting; they have taken the process of engagement and valuation with health issues to far greater depths.

Over their 25-year history, they have promoted a legacy of investment in GI in and around former industrial cities to address health and well-being issues through direct community engagement. They have achieved this by reacting to the needs of local people to re-engage with local environments and investing in projects that promote accessibility, multifunctionality, and social interaction. Investments in GI therefore provide the environmental resources through which behavioral change in attitudes toward the use of local landscapes can be measured. The longer term benefit of this process has been to engender a sense of capacity within people to actively manage their own health through the use of community forest sites.

Furthermore, environments perceived to be of a higher quality have been shown to have a consistent, and long-term, effect on social cohesion and communal well-being as compared to lower quality environments (Mell, 2009; Pretty et al., 2005; Town and Country Planning Association and The Wildlife Trusts, 2012). Consequently, where CFPs have invested in GI resources, there has been a visible and, in most cases, quantifiable increase in the use of the landscape and outdoor activity. This behavioral change has been linked to the increased number of affordances[7] that improved accessibility, location, and the multifunctional nature of GI. It also reflects the location of these resources in areas of greatest need.

This chapter has proposed that GI encourages positive shifts in personal and communal behavior as it provides for short-term and, in many cases, immediate modification of individual health activities as well as supporting longer term changes as people utilize community forest resources more frequently with subsequent improvements in health.

One final thought reflects on the process of reimaging urban forests in the UK. Through a process of reconnection with urban GI, people are beginning to re-evaluate the value of community forests, urban woodlands, and GI in general. Communities have always valued trees, but in an era when obesogenic environments are present, where the value of property is the most frequent economic conversation people hold, and where urban ecology is being marginalized by development, urban trees and forests remain one of the few environmental resources associated with positive memories (Louv, 2005). Such a collective positivity has been engaged by the CFPs that have successfully targeted some of the most important social issues in England (childhood obesity, economic recession, social exclusion, and a lack of urban greening) often missing in environmental project work. Community Forest Partnerships in England can thus be considered to have made significant and long-term modifications to the physical landscape which have led to changes in personal and communal approaches to health and well-being. By providing the physical resources and the project support, they have encouraged a change in the attitude of people toward the value of GI for health.

Notes

1. Figure 11.2 is differentiated with the darker green highlighting the community forests which are still operational. The lighter green areas illustrate those

partnerships that have either ceased trading or that have been subsumed by other agencies. Some forests represented by light green were created after the original 12 but have since ceased trading or have been subsumed.

2. At the lower super output area level.

3. The data used in this chapter were provided through personal communications with offices and former staff at the Red Rose Forest, The Mersey Forest, and the North East CFP.

4. Both ceased trading in 2008.

5. This includes "The work of the group contributes to Biodiversity and Habitat and Species Action Plans" and "The broader community appreciates the work our Green Gym group does" (Yerrell, 2008).

6. The project reports developed for the VALUE projects in Manchester and Sheffield highlight the responses of *ex-ante* focus group discussions and *ex-post* large-scale surveys in both locations. In both locations the people were asked to discuss the value of green infrastructure to their lives and reported that a "greener" environment made their homes/communities more liveable. See Mell, Keskin, Hehl-Lange, and Henneberry (2012a, 2012b) for a more in-depth explanation of this research.

7. Affordances are the properties of a resource that allow a person or a community to utilize it for an activity. Louv (2005) discusses the value of environmental affordances as a central process in developing a long-term relationship between the environment and personal behavior to address Nature Deficit Disorder.

References

Blackman, D., & Thackray, R. (2007). *The green infrastructure of sustainable communities*. North Allerton, UK: England's Community Forest Partnership.

Brown, C., & Grant, M. (2005). Biodiversity & human health: What role for nature in healthy urban planning? *Built Environment, 31*(4), 326–38.

Coles, R. W., & Bussey, S. C. (2000). Urban forest landscapes in the UK—progressing the social agenda. *Landscape and Urban Planning, 52*(2), 181–8.

Davies, C., Macfarlane, R., McGloin, C., & Roe, M. (2006). *Green infrastructure planning guide*. Annfield Plain, UK: North East Community Forest.

Department of Health. (2004). *Choosing health? Choosing activity: A consultation on how to increase physical activity*. London: HMSO/Department of Health.

DTLR (Department for Transport, Local Government and the Regions). (2002).

Green spaces, better places. Final report of the Urban Green Spaces Taskforce. London: Department for Transport, Local Government and the Regions.

England's Community Forests. (2004). *Quality of place, quality of life.* Newcastle: England's Community Forests.

England's Community Forests & Forestry Commission. (2012). *Benefits to health and wellbeing of trees and green spaces.* Farnham, UK: England's Community Forests & Forestry Commission.

Forest Research. (n.d.). *Health and well-being: The role of woodlands in the North East of England.* Farnham, UK: Forest Research.

Forest Research. (2006). *The Chopwell Wood health project research summary.* Farnham, UK: Forest Research.

Forest Research. (2010). *Benefits of green infrastructure.* Farnham, UK: Forest Research.

Gateshead Council. (2005). *A neighbourhood plan for the Orthodox Jewish community of Gateshead.* Gateshead, UK: Gateshead Council.

Green Infrastructure North West. (2010). *Liverpool green infrastructure strategy.* Liverpool: Green Infrastructure North West.

Kitchen, L., Marsden, T., & Milbourne, P. (2006). Community forests and regeneration in post-industrial landscapes. *Geoforum, 37*(5), 831–43.

Land Use Consultants with SQW Ltd. (2005). *Evaluation of the community forest programme.* London: Land Use Consultants.

Layard, R., Clark, D., Knapp, M., & Mayraz, G. (2007). Cost–benefit analysis of psychological therapy. *National Institute Economic Review, 202*(1), 90–8.

Liverpool City-Region Health is Wealth Commission. (2008). *Health is wealth.* Liverpool: Liverpool City-Region Health is Wealth Commission.

Louv, R. (2005). *Last child in the woods: Saving our children from nature-deficit disorder.* Chapel Hill, NC: Algonquin Books.

Lovasi, G. S., Quinn, J. W., Neckerman, K. M., Perzanowski, M. S., & Rundle, A. (2008). Children living in areas with more street trees have lower prevalence of asthma. *Journal of Epidemiology and Community Health, 62*(7), 647–9.

Maas, J., Verheij, R. A., Groenewegen, P. P., de Vries, S., & Spreeuwenberg, P. (2006). Green space, urbanity, and health: How strong is the relation? *Journal of Epidemiology and Community Health, 60*(7), 587–92.

McDermott, M., & Schreckenberg, K. (2009). Equity in community forestry: Insights from North and South. *International Forestry Review, 11*(2), 157–70.

Mell, I. C. (2006). *North East Community Forest green infrastructure project: Green infrastructure & social exclusuion. Social exclusion at Herrington and Hetton Lyons Country Parks.* Annfield Plain, UK: North East Community Forest.

Mell, I. C. (2007). Green infrastructure planning: What are the costs for health and well-being? *International Journal of Environmental, Cultural, Economic and Social Sustainability, 3*(5), 117–24.

Mell, I. C. (2009). Can green infrastructure promote urban sustainability? *Proceedings of the ICE—Engineering Sustainability, 162*(1), 23–34.

Mell, I. C. (2011). The changing focus of England's Community Forest programme and its use of a green infrastructure approach to multi-functional landscape planning. *International Journal of Sustainable Society, 3*(4), 431–46.

Mell, I. C., Keskin, B., Hehl-Lange, S., & Henneberry, J. (2012a). *Action 4.2 case study report—Street tree investments on Whitworth Street West, Manchester.* Sheffield: University of Sheffield.

Mell, I. C., Keskin, B., Hehl-Lange, S., & Henneberry, J. (2012b). *Action 4.2 level II report: A contingent valuation of green investments in the Wicker Riverside, Sheffield.* Sheffield: University of Sheffield.

Mell, I. C., Henneberry, J., Hehl-Lange, S., & Keskin, B. (2013). Promoting urban greening: Valuing the development of green infrastructure investments in the urban core of Manchester, UK. *Urban Forestry & Urban Greening, 12*(3), 296–306.

Mersey Forest. (2013a). *Summary of natural choices for health & wellbeing evaluation report.* Risley Moss, UK: Mersey Forest/NHS.

Mersey Forest. (2013b). *The Mersey Forest plan: Final draft, September 2013.* Risley Moss, UK: Mersey Forest.

Natural Economy Northwest. (2008). *The economic value of green infrastructure.* Kendal, UK: Natural Economy Northwest.

Natural England. (2011). Walk yourself to a healthier lifestyle. Retrieved May 31, 2015, from http://webarchive.nationalarchives.gov.uk/20140605090108/http://www.naturalengland.org.uk/regions/north_east/press_releases/2011/180111.aspx

O'Brien, E. (2005). *Trees and woodlands: Nature's health service.* Farnham, UK: Forest Research.

O'Brien, E., & Snowdon, H. (2007). Health and well-being in woodlands: A case study of the Chopwell Wood Health Project. *Arboricultural Journal, 30*, 45–60.

Pretty, J., Peacock, J., Sellens, M., & Griffin, M. (2005). The mental and physical health outcomes of green exercise. *International Journal of Environmental Health Research, 15*(5), 319–37.

Pretty, J., Peacock, J., Hine, R., Sellens, M., South, N., & Griffin, M. (2007). Green exercise in the UK countryside: Effects on health and psychological well-being, and implications for policy and planning. *Journal of Environmental Planning and Management, 50*(2), 211–31.

Red Rose Forest. (2013). Red Rose Forest green streets case study: Meanwhile food growing site, Manchester. Retrieved May 31, 2015, from www.redrose forest.co.uk/web/images/stories/macdonalds case study1.pdf

Red Rose Forest. (2014). Retrieved May 31, 2015, from www.redroseforest. co.uk/web/

Ridgers, N. D., & Sayers, J. (2010). *Natural play in the forest: Forest school evaluation (children).* Liverpool: Liverpool John Moores University/Mersey Forest.

Snowdon, H. (2005). *Evaluation of the Chopwell Wood health project.* Newcastle: Northumbria University.

South Yorkshire Forest Partnership & Sheffield City Council. (2012). *The VALUE project: The final report.* Sheffield: South Yorkshire Forest Partnership & Sheffield City Council.

TEP. (2005). *Advancing the delivery of green infrastructure: Targeting issues in England's northwest.* Warrington, UK: The Environment Partnership.

Town and Country Planning Association. (2012). *Creating Garden Cities and suburbs today: Policies, practices, partnerships and model approaches.* London: Town and Country Planning Association.

Town and Country Planning Association and The Wildlife Trusts. (2012). *Planning for a healthy environment—good practice guidance for green infrastructure.* London: Town and Country Planning Association and The Wildlife Trusts.

Ulrich, R. S. (1984). View through a window may influence recovery from surgery. *Science, 224*(4647), 420–1.

Urban Task Force. (1999). *Towards a strong urban renaissance.* London: HM Stationery Office.

Welk, G. J. (1999). The Youth Physical Activity Promotion model: A conceptual bridge between theory and practice. *Quest, 51,* 5–23.

Yerrell, P. (2008). *National evaluation of BTCV's Green Gym.* Oxford: Oxford Brookes University.

Chapter Twelve

Valencia's Jardín del Turia Park

From natural disaster to valued public health amenity

Salvador del Saz Salazar

Valencia, founded by the Romans in 138 BCE, is a Mediterranean city on the east coast of Spain. With over 1.5 million residents in its metro area, it is the third most populated city in Spain after Madrid and Barcelona. The most densely populated of its 19 districts (Figure 12.1) are the Cuitat Vella, the historic core of the city, and the districts that surround it. Over its 2,000-plus years of history, Valencia has expanded outwards from the original Roman settlement along the winding west to east course of the Turia river. In Figure 12.1, the old course of the river forms the northern border of districts 7, 3, 1, 2, and 10 on its way to the sea. This is the "old" course of the river because the Turia no longer flows here. This river corridor, once a maritime artery of trade, is now the Jardín del Turia Park, and how it was formed is a story of how a social tragedy spurred the creation of a world-class urban GI amenity.

The Jardín del Turia Park: the backbone of Valencia

On October 13, 1957, the upstream towns of Lliria, Villmarxant, and Chelva received 500mm of rainfall over a two-day period which caused the typically benign Turia river running through Valencia to rise. As it did so, logs and debris carried by the rising water began blocking the bridges that connected the north and south banks of city. Soon after midnight, more than 1,000 cubic meters of

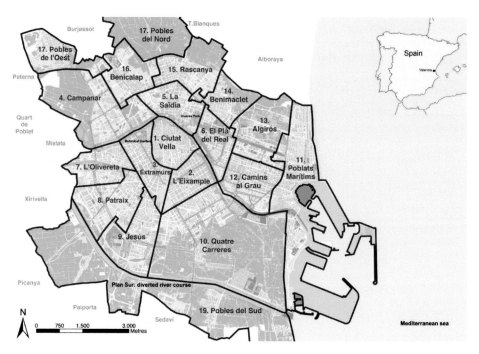

Figure 12.1: Valencia city districts
Source: Salvador del Saz Salazar.

water per second flowed into the streets, reaching over two meters in depth in some areas. At 4am, the flood reached its peak of approximately 2,700 cubic meters of water per second, but then quickly receded. Valencians thought the worst was over, so they ventured out into the water and mud-filled streets without realizing that the worst was yet to come. The upstream towns had been inundated with a second deluge. A wall of water hit Valencia again around 2pm, this time with a flow of 3,500 cubic meters per second. Making matters worse, a heavy rain of around 100mm fell in just half an hour on Valencia, something never before witnessed. The combination of rainfall upstream and in Valencia caused the water to reach two to five meters in depth in some places of the city covering 2,200 hectares. Many houses and buildings collapsed and 81 lives were lost, although the real figure was likely much higher.

As a consequence of this catastrophe, a plan to divert the river was initiated in 1965 and completed in 1972. The "Plan Sur" included the construction of a new riverbed south of the city with a capacity of 5,000 cubic meters per second, a length of 11.8 kilometers, and a width of 175 meters. Figure 12.1 shows the Plan Sur river channel on the southern edge of the city.

When the Plan Sur was completed, the question was then what to do with the old and now completely dry riverbed. Two alternatives were considered, a highway and a park. Strongly influenced by the prevailing growth-at-all-costs development vision of the late 1960s and early 1970s, the construction of a multilane highway would have created a connection between landlocked Madrid and the Port of Valencia. This alternative was rejected largely because of the backlash it received from Valencians with many arguments revolving around how the traffic speeding through the heart of the ancient city would negatively impact many aspects of their well-being. The oil crisis of the 1970s also contributed to the eventual rejection of this alternative. The park creation camp won out, and in 1979 the old riverbed was designated as a greenspace. Construction began shortly thereafter, and the first three sections of the proposed 18 sections of the park were inaugurated in 1986. The remaining sections were gradually constructed without a clear guiding thread, thus resulting in a "collage" of design features and uses.

It cannot be overlooked that the creation and construction of the park were influenced by the complexities of the Spanish transition to democracy. In fact, the perceived need for conserving the riverbed as greenspace made it a focal point for a newfound ability to express social demands as well as opinions on how the park should be designed (Sorribes, 2010). For example, some sections were the subject of controversy due to the excessive use of concrete as opposed to natural features. Turning the social tragedy of the 1957 flood into a social victory, the world-class urban park that is today the Jardín del Turia is a natural monument to the past and one that contributes to a sustainable future that promotes health and well-being. The Jardín del Turia has been such a success that it has instigated a continued tradition of expanding greenspace throughout the city.

Availability of greenspace

The amount of public greenspace in Valencia has been growing steadily over the past two decades, increasing 1.67 fold since 1996 (Table 12.1). While this is a positive trend, more telling is whether this amount of greenspace is meeting the needs of a simultaneously growing population. Examining the amount of greenspace per person in Valencia, the results are slightly less favorable. The increase from $4.11m^2$ of greenspace per person to $6.44m^2$ is an increase of a factor

Table 12.1: Amount of greenspace in Valencia

Year	*Total greenspace (m²)*	*Greenspace (m² per person)*
1996	3,074,493	4.11
2000	3,689,448	5.00
2005	4,292,763	5.33
2010	4,810,434	5.94
2012	5,135,126	6.44

Source: Municipal Statistics Office, Valencia city hall.

of 1.56. This ratio falls considerably short of the general target recommended by the World Health Organization (WHO) of a minimum of 9m² of greenspace per city dweller (Kuchelmeister, 1998). Comparing Valencia to others large cities in Spain, the only other city with a comparable amount of greenspace per person is Barcelona (6.4m²). Other major cities such as Madrid and Seville are well over 20m², but the accessibility created by the corridor design of Valencia's most prominent greenspace might very well make up for it falling short of WHO goals.

Characteristics of the park and park users

From west to east, the Jardín del Turia Park traverses the city for 12 kilometers with an average width of 160 meters. The park is divided in 18 sections that differ somewhat in their landscape design and in the facilities that they offer users (Figure 12.2). The park is crossed by 18 bridges constructed between the fifteenth and the twenty-first centuries of which two are exclusively for pedestrian use. Just as the Turia river used to flow under these bridges, they now allow users of the park to pass under the roadways above uninterrupted by potential conflicts with vehicles over its 12km length.

Although the park could be considered a collage of many sections of distinct character, many adjacent sections share similar design and facilities and can therefore be grouped together for the purpose of analyzing the relationship between park characteristics and user characteristics. Sections 1 to 3 contain playground equipment for children, an outdoor running track, a soccer facility, a BMX track (Figure 12.3).

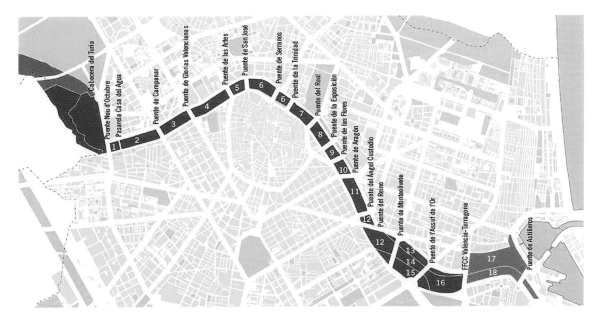

Figure 12.2: Sections of Jardín del Turia Park

Source: Área de Medio Ambiente y Desarrollo Sostenible, Delegación de Parques y Jardines, Ayuntamiento de Valencia.

Figure 12.3: Section 3: Running track

Source: Salvador del Saz Salazar.

Sections 4 to 6 offer a different assortment of sport facilities for baseball, rugby, soccer, and skating in addition to playground equipment (Figures 12.4 and 12.5).

Figure 12.4: Section 6: Jardín del Turia Park as viewed from atop the Torres de Serrano looking west

Source: Salvador del Saz Salazar.

Figure 12.5: Section 6: Jardín del Turia Park as viewed from atop the Torres de Serrano looking east

Source: Salvador del Saz Salazar.

All sections of the park contain natural features, but sections 4 to 6 reproduce a natural Mediterranean forest with a high density of trees. Sections 7 and 8 create a transition area with outdoor sculptures and also a soccer facility. Section 9 is a multiuse area that is used mainly for firework shows and outdoor fairs. Sections 10 and 11 are characterized by two ponds and a music water fountain just opposite the Palau de la Mùsica (Concert Hall) (Figure 12.6).

Section 12 is dedicated to amenities for young people. It holds a unique playground called "Gulliver Park" with an enormous fiberglass recreation of Jonathan Swift's character bound to the ground (Figure 12.7).

The sweeping curves of this giant sculpture create a natural climbing surface for children, and hidden steps provide quick access to multiple levels of ropes, ladders, slides. This section also holds a skateboarding and bike facility that is very popular among young people (Figure 12.8).

Finally, sections 13 to 16 contain the world-renowned City of Arts and Sciences (CAS) made up of an Opera House, Science Museum (Figure 12.9), an IMAX

Figure 12.6: Section 11: In front of the Palau de la Mùsica
Source: Salvador del Saz Salazar.

275

Figure 12.7: Section 12: Gulliver playground
Source: Salvador del Saz Salazar.

Figure 12.8: Section 12: Skate park with Palau de los Artes in the background (left)
Source: Salvador del Saz Salazar.

Theater, a building for hosting a variety of events (The Agora), and an aquarium. This area also has playgrounds for children and several water features.

To the east are sections 17 and 18, still in progress and currently inaccessible, that will eventually connect the park with the port area and the sea.

A survey of 1,480 park visitors conducted *in situ* (Saz Salazar & Rausell-Köster, 2008) revealed that the vast majority of park users live in Valencia (86.2 percent). A slight exception occurs in the sections where the CAS complex is located. Here there are slightly more non-residents (21.4 percent), but this was expected since the CAS is a popular tourist attraction.

The kinds of activities visitors perform in the park depend on the facilities provided in each section. As shown in Table 12.2, in general, the main reasons for visiting the park are strolling (55.4 percent) and playing sports (19.6 percent). Strolling occurs somewhat consistently across all sections of the park. Across all sections there are numerous paths with dirt and paved surfaces. Walking and organized sports provide physical health benefits, mental health benefits from

Figure 12.9: Section 15: Museo de las Ciencias
Source: Salvador del Saz Salazar.

exposure to nature, and the potential for social interaction. Unfortunately, there was no accounting of cycling and running in this study even though these are common activities in the park. This was due to the difficult nature of stopping people performing these activities in order to collect survey data. Biking *is* captured when measuring accessibility below. Playing sports is performed in sections where the concentration of sports facilities is highest (1–3, 7–8). Gulliver Park playground, section 12, instigates the highest level of playground activity (13.5 percent).

Knowing that age is an important determinant of the types of activities performed, Table 12.3 shows the activities by age group. Three apparent and expected trends emerge. First, in general, the older the visitor, the more leisure strolling activity. We also see a spike in those 56–65 walking the dog. Second, the younger the visitor the more likely they are to use the park for sports. Having facilities to support both walking and sports is essential to provide potential health benefits to a wide array of ages of varied preferences and capabilities. The availability of the park to meet friends is also an important feature for young people. Third, the youngest visitors (26.3 percent) and their parents (aged 26–45) are found in the playground facilities dispersed throughout different sections of the park.

The accessibility of the park was also examined. The European Environment Agency recommends that people should have access to greenspace within 15-minute walking distance, while in the UK the recommendation is 300 meters from home (Barbosa et al., 2007). Considering an average walking speed of 5km/h, the 300 meter mark makes the time distance to reach a park 3.6 minutes. At the same walking speed, one could cover 1.2km in 15 minutes. There is a large discrepancy between a minimum of 300m and 1.2km from greenspace. A compromise was struck by using the 10-minute rule of thumb distance commonly used in greenspace accessibility studies (Coutts, 2008). Only 29 percent of visitors to the Jardín del Turia were within a 10-minute walking distance, while those between a 10-minute and 30-minute walking distance accounted for 53 percent of visitors. Among those on a bike and considering an average bike speed of 12km/h, 24.1 percent were within a 10-minute bike distance (2km) and 57.1 percent were within a 10-minute and 30-minute biking distance (6km). These results reveal that most park visitors are willing to travel further than the 10-minute walking and biking distance to access the park. This speaks to the attractive effect of its ability to support a wide range of activities and also, likely, its size. Once reached, the park

Table 12.2: Activities by sections of the park (percentages of total)

Activity	1–3	4–6	7–8	9	10–11	12	13–16	All
					Sections			
Strolling	45.2	53.6	48.5	52.1	57.5	44.1	64.0	55.4
Playing sports	37.6	20.0	24.3	19.6	18.8	19.8	14.5	19.6
Reading	5.4	3.7	9.7	3.7	4.8	9.9	4.8	4.6
Playground	1.1	5.3	3.9	8.6	5.0	13.5	4.8	5.9
Walking the dog	1.1	4.2	3.9	4.3	3.8	4.5	1.0	3.6
Meeting friends	4.3	6.3	9.7	6.7	5.8	3.6	4.2	5.7
Other activities	5.3	6.9	0.0	5.0	4.3	4.6	6.7	5.2

Source: Salvador del Saz Salazar.

Table 12.3: Activities by age groups (percentages of total)

Activity	0–15	16–25	26–35	36–45	46–55	56–65	>65
			Age in years				
Strolling	31.6	40.8	52.0	58.3	81.6	72.5	75.2
Playing sports	26.3	30.7	24.4	15.4	5.9	3.8	2.1
Reading	0.0	4.1	6.0	4.3	1.5	5.0	7.1
Playground	26.3	1.5	7.4	10.6	4.4	5.0	6.4
Walking the dog	5.3	2.4	2.8	5.1	1.5	10.0	4.3
Meeting friends	10.5	11.4	3.7	2.4	2.9	1.3	1.4
Other activities	0.0	9.1	3.7	3.9	2.2	2.4	3.5

Source: Salvador del Saz Salazar.

user can travel uninterrupted through the park, without having to cross a road, for 12 kilometers (and over triple this distance considering the connection of the park to further greenspace to the west that will be discussed shortly).

The attractiveness is revealed again when considering the amount of time spent in the park. A majority (60.7 percent) of visitors stated they spent between 30 to 90 minutes in the park, with an additional 17.2 percent spending more than 90 minutes. Considering that almost all visitors were performing some type of physical activity and assuming the activity is being performed during most of their park visit, this is more than enough time to meet the recommended level of activity to achieve health benefits, not to mention the other health benefits that are achieved through exposure to nature and social interactions. So, what value do Valencians place on the park?

A monetary approach to the social benefits of the Jardín del Turia

The many ecosystem services of green infrastructure (GI) have been outlined in previous chapters with the focus being on how these services provide human health benefits. While people may not articulate the benefits they receive from GI as "ecosystem services," studies cited in Chapter 3 reveal that people do attribute greenspace with physical and mental health benefits. These health benefits are captured when people are asked to translate the value of these services into monetary terms. Considering the non-market nature of the numerous services that parks provide, their monetary value is captured using stated-preference methods such as the contingent valuation method (CVM) (Mitchell & Carson, 1989). The CVM is a technique that involves the direct questioning of people to elicit their willingness to pay (WTP) for a particular good or service, implying an improvement in their well-being. In the same study that examined the characteristics of users of the Jardín del Turia, Saz Salazar and Rausell-Köster (2008) also conducted a contingent valuation survey aimed at estimating the social benefits derived by Valencians from the use of the Jardín del Turia. They found that the average park user would be willing to pay €7.6 annually in extra property taxes for the array of benefits that they believe they receive from the park. The WTP varied depending on which section of the park the interview was conducted in. This result was apparently explained by the different facilities offered by each section and also the income level of the adjacent areas (Table 12.4). For example, sections 1 to 3 and 4 to 6 exhibited a higher WTP since they offer the visitor various sports facilities and a typical Mediterranean forest for strolling. Conversely, in sections 13 to 16, where the City of Arts and Sciences complex (CAS) is located, the low value obtained (€4.74) could be attributed to people's unwillingness to pay any more than the fees already associated with accessing many of the CAS attractions. They also found that the higher the respondent's income and education, the higher his/her WTP. In sections 10 to 11 (Palau de la Mùsica area), the high WTP (€9.00) is explained by the high income of the adjacent areas. WTP decreased with the number of visits to the park as a consequence of the decreasing marginal utility of each additional visit. Finally, Saz Salazar and Rausell-Köster (2008) conclude

Table 12.4: Visitors' WTP by sections of the park

Sections	*Mean WTP (€)*
1–3	10.50
4–6	10.45
7–8	5.84
9	8.80
10–11	9.00
12	2.66
13–16	4.74
All	7.60

Source: Salvador del Saz Salazar.

that assuming a conservative estimate of the useful life for the park at 25 years, the present value of the expected social benefits would amount from a minimum value of €38.1 million at a 3 percent discount rate to a maximum value of €61.7 million applying a 1 percent discount rate. The enormous value that people place on the park has spurred the extension of the greenspace along the river corridor.

Extending Valencia's green infrastructure system

The success of the Jardín del Turia—represented in its extensive use by a wide cross-section of Valencians and their willingness to pay for the benefits it provides—has spurred a series of extensions to the park which have resulted in this urban greenspace now becoming a part of a regional green corridor. In 2004, the Parque de Cabecera extended the western end of the Jardín del Turia to where the old and new river channels diverge adding to it an additional 330,714m^2 of greenspace. A few years later, in 2007, the regional government created the Fluvial Park of the Turia River that extends from the Parque de Cabecera 35 kilometers upstream along the current and original course of the Turia river. The Fluvial Park has largely conserved the natural features of the riparian habitat and created access with a multiuse trail heavily used by cyclists (Figure 12.10). Also contained within the Fluvial Park are numerous small farms and gardens (Figure 12.11).

There are also plans to create better accessibility between the urban sections of the Jardín del Turia and other greenspaces in the city, such as the Botanical

Garden and other greenspace hubs, that are adjacent to the old riverbed. So, the river that devastated the city of Valencia in the flood of 1957 has now become an uninterrupted 45-kilometer long band of greenspace and one of its greatest assets in protecting and promoting health.

While there have been great successes in extending Valencia's GI system to the west and into the countryside, completing the final segments (sections 17 and 18) to the east, connecting the Jardín del Turia to the sea, has been marred by roadblocks. First, there is the physical barrier of a railway, but this could be overcome with an additional pedestrian underpass or overpass similar to those that exist in dozens of other places in the park. A much larger obstacle is resistance by the port authority and the neighboring municipality of Nazaret which control this area. As a consequence of the expansion of the port 30 years ago, Nazaret lost two kilometers of beach, and they have thus far resisted any further

Figure 12.10: Fluvial Park pedestrian/bike path
Source: Christopher Coutts.

Figure 12.11: Fluvial Park farm
Source: Christopher Coutts.

perceived encroachment. This is despite the fact that the waterfront area the Jardín del Turia would connect to could benefit greatly from increased access and use. Further complicating the issue, the port area that these final sections would lead to was the target of a massive and controversial redevelopment project undertaken in order to accommodate the 32nd America's Cup in 2003. The Marina Real Juan Carlos the First was of such perceived importance to those in power at the time, and their belief that it would attract foreign visitors and dollars, that it came at the expense of the interests of local residents who advocated for a waterfront with public open space. It was a very controversial issue with protests against it coming from environmental and community groups that claimed that such events did not address the real problems of Valencians, especially those living close to the port area. While hosting this event was a short-term success in terms of foreign visitors and media coverage, after its conclusion the number of visits to this renewed waterfront plummeted. Several attempts have been made to bring life to this area with outdoor music concerts and an open-air cinema, but none has succeeded, and the facilities and infrastructure still remain largely abandoned. The connection of the Jardín del Turia to this area could bring with it much needed activity, but old wounds have thus far prevented this from happening. Water features, blue infrastructure, continually surface as an important health-enhancing complement to GI. This final connection has great potential to spur revitalization and economic development, but great care will need to be taken to ensure that subsequent improvements to well-being reach those that need it most.

The overall success of the Jardín del Turia has also spurred another ambitious plan to reclaim greenspace on the south side of the city. The Parque Central plan, signed off in 2003, will take the place of the Estación del Norte railway station that the city plans to relocate. The 230,000m² Parque Central will be structured around a network of pathways aimed at creating different spaces with different functions, such as the Perfume Garden specially designed for visually impaired people and the open-air auditorium for holding cultural events. Parallel to these pathways, there will be eight water channels running through the park and flowing into lakes and fountains. The park will also enhance the current bicycle path network of the city.

The Parque Central will increase the availability of greenspace per inhabitant from the current 6.44m² per person to 6.73m² per person. A modest increase overall, but this brings a much needed greenspace to the districts of the city near the proposed Parque Central (Extramurs and L'Eixample). In these districts, the amount of greenspace per inhabitant will increase from the current 1.23m² per person to 3.65m² per person. This still falls short of the average amount of greenspace for the entire city and the amount recommended by the WHO, but Valencia is on the march to reach this goal.

Summary

The flood of 1957, despite its tragic consequences, triggered an ambitious plan for transforming Valencia's GI. The plan that diverted the Turia river opened the door for a radical transformation of the old riverbed. Public pressure prevented it from becoming a highway, and today it is the health-enhancing green backbone of Valencia. The Jardín del Turia is a public space where people from different social classes interact and reap the health-related benefits associated with exposure and access to GI. The social success of the Jardín has spurred the conservation and reclamation of other greenspaces throughout the city that now extend contiguously into the countryside. Many cities are located on rivers, and the lesson of Valencia is not that cities need to divert rivers to create GI. Riverfront redevelopment, not necessarily diversion, that includes GI and pedestrian access has the potential to provide many health-enhancing benefits.

References

Barbosa, O., Tratalos, J. A., Armsworth, P. R., Davies, R. G., Fuller, R. A., Johnson, P., & Gaston, K. J. (2007). Who benefits from access to green space? A case study from Sheffield, UK. *Landscape and Urban Planning, 83*, 187–95.

Coutts, C. (2008). Greenway accessibility and physical-activity behavior. *Environment and Planning B: Planning and Design, 35*(3), 552–63.

Kuchelmeister, G. (1998). *Urban forestry: Present situation and prospects in the Asia and Pacific region.* Rome: Food and Agriculture Organization of the United Nations.

Mitchell, R. C., & Carson, R. T. (1989). *Using surveys to value public goods: The contingent valuation method.* Washington, DC: Resources for the Future.

Saz Salazar, S., & Rausell-Köster, P. (2008). A double-hurdle model of urban green areas valuation: Dealing with zero responses. *Landscape and Urban Planning, 3*, 241–51.

Sorribes, J. (2010). Valencia: La huerta, el río y el mar. In J. Sorribes (Ed.), *Valencia, 1957–2007: De la riada a la Copa del América* (pp. 15–35). Valencia, Spain: Publicacions de la Universitat de València.

Chapter Thirteen

Health and Hamburg's *Grünes Netz* (Green Network) Plan

Thomas B. Fischer

In early 2014, the city of Hamburg, Germany, made headlines with a development plan that allegedly aimed to eliminate cars from its streets in 20 years (Quirk, 2014; Paterson, 2014), replacing auto transport with walking and cycling. Addressing whether this was possible, the British Broadcasting Corporation reported that "city officials obviously feel that the personal motorcar does not fulfill a function that walking, biking and taking public transport cannot" (Stewart, 2014). In health terms, such a bold step would mean a reduction in the 10,000 injured and over 30 human deaths per year caused by pedestrian and bicycle encounters with vehicles in the city (Statistisches Bundesamt, 2013). There would also be other noted positive mental and physical health effects associated with increased levels of physical activity. So, how did Hamburg plan to achieve this? At the heart of the headline-grabbing plan was a *Grünes Netz* (green network) of interconnected open areas that would cover approximately 40 percent of the city. This network complemented sidewalks and bike lanes on roadways with a separate green transportation system dedicated to walking and cycling modes.

Rightfully so, legitimate questions were raised as to whether such an ambitious plan was realistic in a port city of trade with nearly two million inhabitants. In the months following the headlines, corrections and clarifications began to emerge. There had never been any plans to fully eliminate cars in Hamburg.

Germany's second-largest city does not want to go 'car-free' within two decades, as many media reports wrongly stated earlier this year, it is weaving a '*Grünes Netz*'—a green network of parks, playgrounds, sports fields, bike paths and the like which will allow pedestrians or cyclists to more easily navigate through the city. In other words, cars won't be banned, but get downgraded—a big deal in a country which loves its cars almost as much as America.

(Anonymous, 2014)

While this clarification came as a disappointment to some, it was clear that Hamburg still had ambitious plans to substantially increase walking and cycling. It also spurred enormous interest in the green network at the heart of the plan.

This chapter elaborates on Hamburg's Green Network Development Plan (GNDP), the explicit place of health in the plan, and how the *Grünes Netz* is essential to other plans and programs that also aim to improve health. First, Hamburg's plan is placed in the context of green infrastructure (GI) development in Europe and the German landscape planning system where the prominence of health is increasing. For those with a planning inclination, the case of Hamburg reveals how health is an important undercurrent in landscape planning and a potentially powerful motive that, if brought to the fore, can link the environmental and social goals of other plans and administrative authorities. Likewise, for those with a primary focus on health, this case provides some guidance on how planning is done and how landscape protection and health promotion are inseparable.

Green infrastructure development and human health in Europe

Over the past two decades in Europe, there has been a renewed interest among spatial and development planners in the conservation and reintroduction of GI in built-up areas. The particular interest on GI in Europe is connected with its perceived ability to deliver a range of benefits, such as the creation of networks of habitats, flood risk reduction, and improvements in human physical and mental health and quality of life. Benefits are frequently explained and operationalized in terms of ecosystem services all of which have numerous, and not mutually exclusive, connections to human health and well-being (World Health Organization, 2005).

Mazza et al. (2011) identified approximately 100 European GI initiatives and found recreation and health benefits to be important objectives in most of them, although they stress that health benefits were difficult to attribute directly to GI (e.g. climate-related health issues). The complexities of isolating the human role within ecosystems and the human health benefits derived from ecosystem services are a challenge, but, as we have seen in Chapter 3 with climate change being a stark example, estimates of health impacts are being made. There is ample scientific evidence revealing that GI is indispensable to the provision of the fundamental ecosystem services of water, food, and air, and GI's role in infectious disease ecology, physical activity, mental restoration, and social capital. We have arrived at a point where it is no longer a matter of *if*, but rather a matter of applying increasingly sophisticated methods to measure *how much*. There is ample evidence to proceed with GI development as a health promotion strategy, and doing so will allow more pre-/post-impacts to be measured. During this process, there is solace in the fact that it is highly unlikely there is such a thing as too much GI; GI conservation is always likely a safe bet for health promotion.

It is important to note that in many European countries the issues and associated debates around GI conservation are not entirely new, with many of them having been important policy and planning considerations for some time. Planning for GI has been a part of most planning systems since they were first developed, either in fully integrative ways (e.g. the UK's Town and Country Planning system) or as a distinct planning process (e.g. the Dutch National Ecological Network; the German Landscape Planning system). Associated planning tools have also been in place for some time in many countries, such as the Green Belt designation. These plans fulfill numerous functions not the least of which is their support of human health and well-being (Matsuoka & Kaplan, 2008).

GI development at many scales is continuing to expand in Europe with an EU-wide GI strategy currently being developed and various EU member states creating trans-border regional plans such as the Latvian, Hungarian, and Czech national ecological networks (European Centre for Nature Conservation, 2014; Jones-Walters, 2007). Health is coming to the fore in some of these plans. At the country level, following the European Green Paper on Green Infrastructure, the German federal government has announced its intention to prepare a Federal Concept for Green Infrastructure (Anonymous, 2013) that includes explicit references to health benefits at various points. In the UK, a number of guidance documents on the health benefits of GI have been prepared (e.g. Town and Country Planning Association, 2012; Natural England, 2009; Spatial Planning and Health Group, 2011). The accumulation of evidence documenting the connection between GI and health has led to administrative silos being broken down.

The National Health Service of the UK is playing an important role in folding into GI plans the substantial benefits to physical and mental health, particularly in urban environments (Houses of Parliament, 2013). In urban environments, such as with Hamburg, health and well-being have been an increasingly important goal of GI conservation (Natural Economy Northwest, 2008).

Landscape planning in Germany: the context for Hamburg's Green Network Plan

The roots of landscape planning in Germany date back to the early twentieth century with statutory requirements for the preparation of landscape plans and programs taking hold under the Federal Nature Conservation Act of 1976. Under the renamed Federal Environmental Protection Act, the protection of the natural landscape is viewed as essential both for its inherent value but also for its recognized role as forming the basis of human life for current and future generations (von Haaren, 2004). In forming the basis of human life, health is at the heart of landscape planning, albeit in an implicit manner. The main aim of landscape plans and programs is to integrate considerations of the environment, nature, biodiversity, and landscape into decision-making and planning so that human life and health can be sustained.

Landscape plans and programs in Germany serve as state of the environment reports which proactively set objectives for environmentally sustainable land use (Hanusch & Fischer, 2011). They include information on:

1. The existing and anticipated status of nature and landscape.
2. The objectives and principles of nature conservation and landscape management, forming an important reference frame for spatial/land-use plans (*Flächennutzungsplan*).
3. The assessment and evaluation of the existing and anticipated status of nature and landscape on the basis of overall aims and principles, including any possible conflicts.
4. The anticipated measures for avoiding, reducing, or eliminating adverse effects of planned activities in spatial/land-use plans on nature and landscape, and protecting, managing, and developing certain parts or components of nature and landscapes, among which the European ecological network Natura 2000 through anticipated development (Federal Ministry for Environment, 2002).

Particularly noteworthy is in points 2 and 4: reference is made to how the landscape plan should work in concert with the spatial/land-use plan. To date, Germany is still the only country in the world with a formalized and comprehensive landscape planning system requiring that landscape plans and programs be prepared in parallel with the statutory spatial/land-use planning (Figure 13.1). Hamburg's GNDP, and *Grünes Netz*, was developed under this landscape planning system, and thus landscape planning is also required to consider the social needs in the spatial/land-use plan. As we will see shortly, this created the space for Hamburg's GNDP to coalesce with the transport goals of the spatial/land-use plan which will carry with them improvements to public health.

Hamburg is somewhat unique in that it is among a small group of German city-states (the other two being Berlin and Bremen). For landscape planning purposes, it is considered as among one of the 16 German states for which a landscape

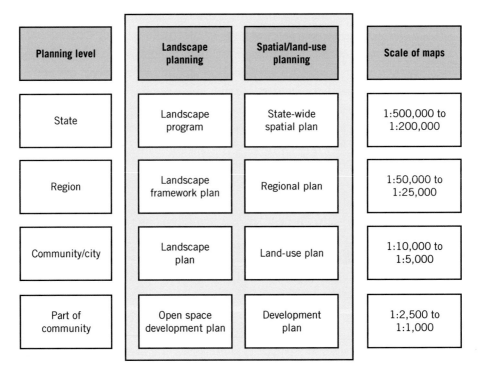

Figure 13.1: System of landscape planning and spatial/land-use planning in Germany

Source: Adapted from Federal Ministry for Environment (1998).

program (*Landschaftsprogramm*) is prepared. However, as it is directly feeding into Hamburg's city-level spatial/land-use plan, it also fulfills the role of a city-level landscape plan. Furthermore, Hamburg's landscape program also acts as a regional landscape framework plan. The level of detail provided in Hamburg's landscape program is close to what is normally expected for a city landscape plan, which would be the primary landscape planning document for Hamburg if it were not a city-state. A focal point of Hamburg's landscape program is the GNDP.

Hamburg's Green Network Development Plan

The city of Hamburg's GNDP is one of the most comprehensive GI plans for a city of over one million inhabitants in Europe, on par with a handful of other European cities such as Stockholm, Copenhagen, Berlin, and Vienna with similarly ambitious plans to conserve GI. The GNDP was critical to Hamburg being awarded the European Green Capital Award in 2011. Launched in 2010, this award has also been granted to other cities including Stockholm (2010), Vitoria-Gasteiz, Spain (2012), Nantes, France (2013), Copenhagen (2014), Bristol, UK (2015), and Ljubljana, Slovenia (2016) (European Commission, 2014).[1]

At the heart of the GNDP is the development of several landscape axes (green corridors) that together form a *Grünes Netz* or open space interconnecting system (Figure 13.2) planned to be fully implemented by 2034. The landscape axes were first laid out in a concept map in 1985 under the landscape program. There are about 10 axes mostly following waterways that lead to the city center and also several smaller axes leading to secondary centers of the city. The plan connects existing green, blue, and other open spaces (e.g. parks and city squares) inside the city to those outside the city. Inside the city, the greenspaces include parks, playgrounds, sports pitches, gardens, and cemeteries. Towards the city edges, axes extend into forests and environmental protection and agricultural areas. The GNDP also complements the objectives of the city-wide open space analysis which aims to formulate green and blue infrastructure standards. This is based on calculations for establishing minimum requirements for features such as playgrounds and city squares to be located in close proximity to where every citizen lives. This network creates the physical GI necessary to achieve a number of interwoven environmental and social goals including supporting and improving the health of Hamburg residents.

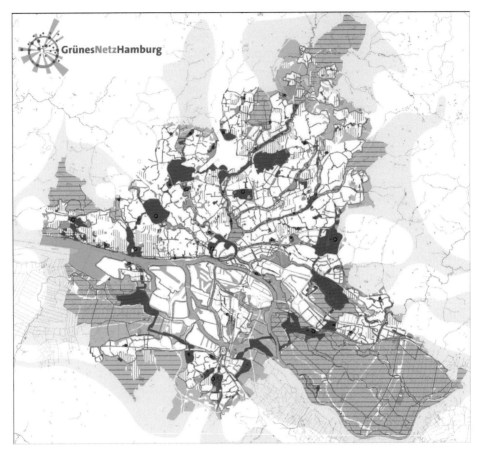

Figure 13.2: Open space interconnecting system
Source: Behörde für Stadtentwicklung und Umwelt (2013).

Health and the GNDP

Landscape planning protects the landscape and its subsequent ability to function and deliver the ecosystem services needed to support human life and health, but the foremost goal of landscape planning is not to promote human health, it is to protect the GI on which health depends. Therefore, there is often no explicit mention of health outcomes in landscape plans even though how well we are planning is evident in the health of the organisms, including humans, that depend on the landscape. Hamburg's GNDP is an exception. The same interconnected system of GI necessary to maintain a healthy landscape is also recognized as promoting health. One of the explicit goals of the GI system that will create recreational and utilitarian walking and cycling corridors is the promotion of physical activity with its resultant physical and mental health benefits (Bayerisches Landesamt für Umwelt, 2014; Behörde für Stadtentwicklung und Umwelt, 2014). Of particular note is the role that GI will play in addressing the worrying issue of childhood obesity.

This inclusion of the social goal of health in the GNDP is in no small part due to the requirement of the landscape plan to work in unison with the spatial/land-use plan where, within the transport element, the physical and mental health benefits associated with promoting cycling are noted. In the spatial/land-use plan, the interconnected open space system in the GNDP is considered as playing a key role in the development of transportation in the city. Green corridors are not just viewed as landscape features critical for landscape functioning but also as pedestrian and cycling corridors that can complement the existing transportation infrastructure (including bike lanes along roads) and cycling strategy (*Radverkehrsstrategie*) of the spatial/land-use plan. In this way, it achieves one of the aims of the spatial/land-use plan cycling strategy which is to reduce physical barriers to cycling. Promoting physical activity, particularly in urban environments, begins with providing the physical infrastructure necessary to perform activity. Wisely, and recognizing the limitations of environmental determinism, the cycling plan also aims to reduce mental barriers by improving public perception through education. With reductions in both physical and mental barriers, the city aims to double the cycling modal split from 9 percent to 18 percent. While this is nowhere near a complete abandonment of the automobile, as the attention-grabbing headlines cited earlier would have us believe, it does double the number of people who would be achieving some level of physical activity. Improving health by improving cycling safety is also an aim. While a skeptical view might lead one to believe that more cyclists will simply mean more people putting themselves at risk, cyclists traveling on a system exclusively dedicated to them will reduce interactions with automobiles. Demonstrating a true commitment, the city of Hamburg aims to invest five million euros per year to finance the cycling network. Because the *Grünes Netz* is such a prominent part of the cycling network, transportation financing is also GI financing.

Landscape planning performed in concert with the goals of the spatial/land-use plan can benefit health, but a hurdle yet to be overcome in Germany is that these plans must remain distinct from health planning. The landscape program is the responsibility of the Authority for City Development and the Environment, and health is the responsibility of the Authority for Social Matters, Family, Health and Consumer Protection (Fischer, Martuzzi, & Nowacki, 2010). As a statutory duty, the former should not take any responsibility for health planning away from the latter. The GNDP is a case where these two administrative silos coalesce. It addresses an identified need of the health authority to increase levels of physical activity. This need has been identified in various health reports, including one on the exercise habits of children (Behörde für Soziales, Familie, Gesundheit und Verbraucherschutz, 2006) and another on elderly people (Behörde für Soziales, Familie, Gesundheit und Verbraucherschutz, 2010). The GNDP supports health planning by providing the physical infrastructure necessary for physical activity and mentions health as one among many reasons for doing so. With landscape and health authorities working together, there is a consistent message: GI is critical to quality of life and health. It may very well be the inseparable nature of landscape and health in plans such as the GDNP that causes the health authorities to revisit GI's essential role as the basis of human life, and, likewise, landscape planning to consider health as a fundamental, but often underemphasized, justification for implementing GI plans.

In creating the physical infrastructure necessary to promote walking and cycling, the GNDP also achieves the goals of another development program with subsequent health benefits beyond those achieved through physical activity but very much dependent on people choosing bikes over their automobiles. The city's environment program (*Umweltprogramm*) outlines the main environmental objectives of the city under three action areas: (1) climate change mitigation and adaptation, (2) sustaining and enhancing quality of life, and (3) developing Hamburg as a green city. The GNDP supports all three of these interdependent action areas. Taking them in reverse order, the GNDP, by definition, contributes to the development of Hamburg as a green city. One among the myriad ways that GI can enhance quality of life is in its ability to address the first action area to mitigate and help humans adapt to climate change. The interconnected open space system that will promote sustainable transportation options will mitigate climate change primarily by reducing auto emissions, but, as outlined in Chapter 3, GI also captures carbon already emitted (dependent on the type of flora), reduces emissions from buildings due to its cooling effect, and can reduce the effects of increases in the frequency and severity of flooding. In Hamburg, a number of developments have been allowed in flood-prone areas (Freie und Hansestadt Hamburg, 1997). As an adaptive strategy, the GNDP also functions as a heat amelioration plan (see e.g. Forest Research, 2010) to reduce the heat island effect in Hamburg. Importantly, in 2011, a City Climate Analysis and Climate Change Scenario (Behörde für Stadtentwicklung und Umwelt, 2012) was produced, and future scenarios are very much a part of future amendments to the city's landscape program. The structure of the GI system will be very much influenced by its recognized role in mitigating climate change and adapting to its effects.

At its core, the GNDP, like planning in general, is aimed at improving people's lives. One of the major ways it does this is by promoting health. The GNDP can be viewed as a health plan in that it creates the physical infrastructure necessary to promote active and sustainable transportation. In doing so it meets the environmental and social goals of the landscape program, the spatial/land-use plan, the health authority, and the environment program. Hamburg has been explicit in connecting active transportation to the health benefits of physical activity and increased safety, but it could truly be a leader if it was to also connect a number of health benefits the *Grünes Netz* will deliver that are currently unstated as co-benefits, such as those related to improved air quality and heat amelioration.

Summary

Hamburg is an award-winning green city with an ambitious GI plan that can be considered among the best practices in landscape planning. The GDNP is not only a plan but a planning paradigm for a city as it supports the environmental and social goals of a number of plans that span the boundaries of administrative authorities. In addition to the environment goals associated with connecting biotopes in the city, an important aim of the GNDP is to ensure a minimum amount of greenspace for the city's population. Doing both is necessary to deliver the ecosystem services on which health depends. The mere presence of GI can ameliorate the negative health consequences of climate change, but access is necessary to support the behaviors (e.g. physical activity) that can enhance health. The GNDP is the backbone of Hamburg's plan to double the levels of cycling and walking in the city with one of the explicit purposes for doing so being improvements to human physical and mental health. By explicitly citing health as a product of GI planning, Hamburg makes GI more than just a luxury; it makes it a recognized necessity for humans to survive and thrive. The GDNP goal of creating an interconnected system of GI is a case where doing right by the environment and humans is one in the same.

Note

1. Awards are handed out years ahead of time so that the award cities can increase their related efforts.

References

Anonymous (2013). BdB: Bundeskonzept grüne infrastruktur "und ausweitung der städtebauförderung wichtige signale der politik. *TASPO*, November 16. Retrieved from http://taspo.de/aktuell/alle-news/detail/beitrag/58637-bdb-bundeskonzept-grune-infrastruktur-und-ausweitung-der-staedtebaufoerderung-wichtige-signale-der-politik.html

Anonymous (2014). Car-free cities: Clearing the air. *The Economist*, March 27. Retrieved from www.economist.com/blogs/schumpeter/2014/03/car-free-cities

Bayerisches Landesamt für Umwelt (2014). Landschaftsplanung—Aufgaben, Ziele und Leistungen. Retrieved from www.lfu.bayern.de/natur/landschaftsplanung/aufgaben_ziele/index.htm

Behörde für Soziales, Familie, Gesundheit und Verbraucherschutz (2006). *Hamburger Kinder in Bewegung*. Retrieved from www.hamburg.de/content blob/116882/data/bewegungsbericht.pdf

Behörde für Soziales, Familie, Gesundheit und Verbraucherschutz (2010). *Die Gesundheit älterer Menschen in Hamburg*. Retrieved from www.hamburg.de/contentblob/2752016/data/pdf-gesundheit-aelterer-menschen-in-hamburg.pdf

Behörde für Stadtentwicklung und Umwelt (2012). *Stadtklimaanalyse und Klimawandelszenario zum Landschaftsprogramm Hamburg*. Retrieved from www.hamburg.de/landschaftsprogramm/3957546/stadtklima-naturhaushalt/

Behörde für Stadtentwicklung und Umwelt (2013). Kartengrundlage: Landesbetrieb Geoinformation und Vermessung, Hamburg.

Behörde für Stadtentwicklung und Umwelt (2014). *Landschaftsachsen innerhalb des 2. Grünen Rings*. Retrieved from www.hamburg.de/landschaftsachsen/3910132/landschaftsachsen-hintergrund/

European Centre for Nature Conservation (2014). *Ecological network maps*. Retrieved from www.ecnc.org/ecological-network-maps/

European Commission (2014). *European green capital award*. Retrieved from http://ec.europa.eu/environment/europeangreencapital/index_en.htm

Federal Ministry for Environment (1998). *Landscape planning—contents and procedures*. Bonn: Federal Ministry for Environment.

Federal Ministry for Environment (2002). *Landscape planning for sustainable municipal development*. Bonn: Federal Ministry for Environment.

Fischer, T. B., Martuzzi, M., & Nowacki, J. (2010). The consideration of health in SEA. *Environmental Impact Assessment Review, 30*(3), 200–10.

Forest Research (2010). *Benefits of green infrastructure*. Retrieved from www.forestry.gov.uk/pdf/urgp_benefits_of_green_infrastructure.pdf/$FILE/urgp_benefits_of_green_infrastructure.pdf

Freie und Hansestadt Hamburg (1997). *Landschaftsprogramm einschliesslich Artenschutzprogramm, Erläuterungsprogramm*. Retrieved from www.hamburg.de/contentblob/3910982/data/erlaeuterungsbericht-landschaftsprogramm.pdf

Hanusch, M., & Fischer, T. B. (2011). SEA and landscape planning. In B. Sadler, R. Aschemann, J. Dusik, T. B. Fischer, M. Partidário, & R. Verheem (Eds.), *Handbook of strategic environmental assessment* (pp. 257–73). London: Earthscan.

Houses of Parliament (2013). *Urban green infrastructure, postnote 448*. Retrieved from www.parliament.uk/briefing-papers/post-pn-448.pdf

Jones-Walters, L. (2007). Pan-European ecological networks. *Journal for Nature Conservation, 15*(4), 262–4.

Matsuoka, R. H., & Kaplan, R. (2008). People needs in the urban landscape: Analysis of landscape and urban planning contributions. *Landscape and Urban Planning, 84*(1), 7–19.

Mazza, L., Bennett, G., De Nocker, L., Gantioler, S., Losarcos, L., Margerison, C., et al. (2011). *Green infrastructure implementation and efficiency*. Final report for the European Commission, DG Environment on Contract ENV.B.2/SER/2010/0059. Brussels and London: Institute for European Environmental Policy. Retrieved from http://ec.europa.eu/environment/europeangreencapital/index_en.htm

Natural Economy Northwest. (2008). *The economic value of green infrastructure*. Kendal, UK.

Natural England (2009). *Green infrastructure guidance, NE 176*. Retrieved from http://publications.naturalengland.org.uk/publication/35033

Paterson, T. (2014). Auto ban: How Hamburg is taking cars off the road. *The Independent,* January 15. Retrieved from www.independent.co.uk/news/world/europe/auto-ban-how-hamburg-is-taking-cars-off-the-road-9062461.html

Quirk, V. (2014). Hamburg's plan to eliminate cars in 20 years. *ArchDaily,* January 7. Retrieved from www.archdaily.com/464394/hamburg-s-plan-to-eliminate-cars-in-20-years/

Spatial Planning and Health Group (2011). Steps to healthy planning: Proposals for action. Retrieved from www.spahg.org.uk/wp-content/uploads/2011/06/SPAHG-Steps-to-Healthy-Planning-Proposals-for-Action.pdf

Statistisches Bundesamt (2013). *Unfallentwicklung auf deutschen Strassen 2012.* Retrieved from https://www.destatis.de/DE/PresseService/Presse/Pressekonferenzen/2013/Unfallentwicklung_2012/begleitheft_Unfallenwicklung_2012.pdf?__blob=publicationFile

Stewart, J. (2014). Can a city really ban cars from its streets? *BBC,* February 4. Retrieved from www.bbc.com/future/story/20140204-can-a-city-really-go-car-free

Town and Country Planning Association (2012). *Planning for a healthy environment—good practice guidance for green infrastructure and biodiversity.* Retrieved from www.tcpa.org.uk/data/files/TCPA_TWT_GI-Biodiversity-Guide.pdf

von Haaren, C. (2004). *Landschaftsplanung.* Stuttgart: UTB.

World Health Organization (2005). *Ecosystems and human well-being—health synthesis.* Retrieved from www.who.int/globalchange/ecosystems/ecosys.pdf

Index

Abiotic, Biotic and Cultural (ABC) services 45n7
access 20–1, 35, 79, 237; Community Forest Partnerships 262, 263, 264; greenways 165; Jardín del Turia Park 278, 284–5; mental health 200; physical activity 162, 168, 173; regular 36; research gaps and needs 238; stress recovery 190; variance by population 240
active travel 169
"adaptation" hypothesis 184
ADHD *see* Attention Deficit Hyperactivity Disorder
affect 179, 182, 190, 191–6, 197, 199, 200; *see also* emotions
affiliation 34, 35, 194
affordances 161, 164, 264, 265n7
Afrane, Y. 152
Africa: biodiversity and disease risk 155; bushmeat 149; deforestation 152; landscape changes 155; savanna 180; water 93
agents 7, 41, 72
aggression 192, 193, 194, 195, 199
agriculture: Community Forest Partnerships 249; deforestation 152; irrigation 93, 153; pest control 111n10;

pollination 103–4; vector-borne diseases 153; *see also* food
air 95–102, 109–10, 240, 288; CHANS 42; co-benefits of exposure to GI 205; dry deposition 110n4; green roofs 134; Health Map 71; Public Health Ecology model 72, 73; valuation of ecosystem services 29, 31, 32
allergens 232, 233, 234n2
Alzheimer's patients 195
animals, threats from 229–30
Antarctica 5
Anthropocene 1–2
anxiety 185, 187–8, 190–1, 192, 200, 202
Arber, S. 216
ART *see* attention restoration theory
Ashton, J. 57
Asia-Pacific 93
Aspinall, P. 198
asthma 32, 72, 98, 99, 232, 261–2
Atlanta 98
attention 180, 181–2, 196–200
Attention Deficit Hyperactivity Disorder (ADHD) 182, 198–9
attention restoration theory (ART) 180, 181–2, 185, 186, 197
Australia 93, 125, 133, 168